Praise for *Being a Human*:

'*Being a Human* is a work of shaggy genius. Its subject is gargantuan in scale; its humour has a reckless panache; its argument is brilliantly original and above all it is written with a matchless audacity of soul. It is one of the most important books I have ever read' Jay Griffiths, author of *Why Rebel?* and *Wild: An Elemental Journey*

'I'll read anything Charles Foster writes, and this is his most ambitious book yet. It is a historical investigation, a short story collection, a humour primer, a sheaf of scientific papers and a work of philosophy all rolled into one, with a side helping of religious ecstasy and badger shit. It will tell you many things you didn't know about who you are. You should read it' Paul Kingsnorth, author of *The Wake*

'*Being a Human* is one of the most original inquiries into the who, what and why of human existence to appear in recent years. Charles Foster writes with inspiring brilliance, originality and simplicity. I love this book. It should be widely read, for the benefit of all us humans' Larry Dossey, author of *One Mind: How Our Individual Mind Is Part of a Greater Consciousness and Why It Matters*

'Monstrously great: book of the year from where I'm sitting. But I'm not sitting, I'm up and waving my arms about for the sustained achievement of this magical, brilliant thing. *Being a Human* contains a hundred things we desperately need to know. Hugely moving, filled with intelligence, it scurries between centuries with us between its teeth. Charles Foster has invoked a living presence in these pages, a contract with the uncanny. To know a thing about the future we need to retrace our steps into our old mind. We could start here' Martin Shaw, author of *Smoke Hole: Looking to the Wild in the Time of the Spyglass*

'A fascinating book of immense scope and proportions ... The evolution of the mind makes for a labyrinthine investigation worthy of Sherlock Holmes' James Crowden, author of *The Frozen River: Seeking Silence in the Himalaya*

'What a mad, brilliant, mind-expanding book. *Being a Human* offers a thrilling deep dive through our evolutionary past, and a witty and learned commentary on why we are the way we are – and what wisdom we've lost along the way. Foster is a true modern polymath who writes with wit, humour and heart: I'll be pressing this book into other people's hands' Cal Flyn, author of *Islands of Abandonment: Life in the Post-Human Landscape*

'Charles Foster has created a book of immense, deeply felt intelligence. This book is a startling reset on our understanding of the journey of human thought. Approaching the question from a totally new perspective of lived experience, Foster shows us how we came to be the people we are, with the values we exert in the world. Not only are the revelations startling, but the metaphoric power of Foster's language is frequently astonishing. I wish I'd written this book, and that's my highest praise' Carl Safina, author of *Becoming Wild: How Animals Learn to be Animals*

'Charles Foster has written the unwritable – gifting us a perspective-tumbling insight into other worlds. *Being a Human* is both challenging and entertaining. By the time you have finished reading it you will not look in the mirror and see quite the same person as before' Hugh Warwick, author of *Linescapes: Remapping and Reconnecting Britain's Fragmented Wildlife*

'Profound, erudite, provocative and funny, this outrageously brilliant and wise book is a challenge to the reductive materialism that dominates current understandings of the human animal – and the natural world. Foster draws on his empathy with the animist Palaeolithic to argue for a return to non-dogmatic forms of Enlightenment values that might take seriously the affective dimension of human nature and experience – to recover "enchantment" and express the "vertiginous wonder of the world" … Wildly eccentric and ranging widely, but always in control' Steve Ely, author of *Englaland*

'Few of us have given much thought to the dazzling human journey from hunter–gatherer to now. In a forty-thousand-year odyssey fizzing with masterful revelation, Professor Foster makes us relive our nature-centric past, shows us how much we have lost and makes us startlingly aware of who we really are' Sir John Lister-Kaye OBE, author of *The Dun Cow Rib: A Very Natural Childhood*

'More turned-down page corners than any other recent book on my shelves. A brilliant, inventive and unsettling exploration of our glorious and broken nature. Foster's work shakes us out of dozy estrangement from our own humanity and welcomes us into the mysteries of belonging … Its richness demands careful reading' David George Haskell, Pulitzer finalist author of *The Forest Unseen: A Year's Watch in Nature*

'A daredevil read. Once again, Charles Foster has journeyed to places most of us wouldn't dare; and emerged with a book that is passionate and kind, deeply intelligent and uproariously funny' Helen Jukes, author of *A Honeybee Heart Has Five Openings: A Year of Keeping Bees*

'Only someone fairly mad – possessed of a sensorial imagination verging on clairvoyance, an alarming appetite for physical duress and an uncanny gift for wyrding his way into other shapes of sentience – would undertake such an impossible endeavour, dropping down and down into the depths within, spelunking in his soul's bone hollows, stirring up old, old ghosts in order to discover how thoroughly haunted our present existence really is' David Abram, author of *Becoming Animal: An Earthly Cosmology*

'No one else could tackle the whole of human evolution, the history and implications of our "inadequate mutations", with such wit and elegance. *Being a Human* is both panoramic and intimate: an experiment in living, a manifesto, a brilliant synthesis, a conversation you'd have in a pub after hours of walking on a wind-scoured moor. Brace yourselves for a thrilling encounter with the other, with the marvellous, terrifying spectacle of the self. This book will leave you changed: both wiser and more bewildered. Which is to say more alive' Helen Mort, author of *Division Street*

'An exhilarating book that asks all the big questions about our past, present and future, *Being a Human* contributes to the growing field of literature that tasks us with thinking, and behaving, like Earthlings. That Foster has managed to produce this clarion call for "a vibrant scientific mysticism" whilst being funny and entertaining is little short of a marvel' Gregory Norminton, author of *The Devil's Highway*

'This is the most wonderful book – deftly written, highly imaginative and a delight to read – and its message is such that its importance simply cannot be overstated. It gives a devastatingly clear portrait of humanity as we have become, and of what we once had – and still could have – but instead are in the process of throwing away, perhaps forever' Iain McGilchrist, author of *The Master and His Emissary*

ALSO BY CHARLES FOSTER

Being a Beast

BEING A HUMAN

*Adventures in
Forty Thousand Years
of Consciousness*

Charles Foster

METROPOLITAN BOOKS

HENRY HOLT AND COMPANY   NEW YORK

Metropolitan Books
Henry Holt and Company
Publishers since 1866
120 Broadway
New York, New York 10271
www.henryholt.com

Metropolitan Books® and ▥® are registered trademarks of
Macmillan Publishing Group, LLC.

Library of Congress Cataloging-in-Publication Data

Names: Foster, Charles, 1962– author.
Title: Being a human : adventures in forty thousand years of consciousness /
    Charles Foster.
Description: First U.S. edition. | New York : Metropolitan Books, 2021. |
    Includes bibliographical references.
Identifiers: LCCN 2021019046 (print) | LCCN 2021019047 (ebook) | ISBN
    9781250783714 (hardcover) | ISBN 9781250783721 (ebook)
Subjects: LCSH: Human behavior—Evolution. | Consciousness—History. |
    Evolutionary psychology. | Human evolution.
Classification: LCC BF698.95 .F67 2021 (print) | LCC BF698.95 (ebook) |
    DDC 155.7—dc23
LC record available at https://lccn.loc.gov/2021019046
LC ebook record available at https://lccn.loc.gov/2021019047

Our books may be purchased in bulk for promotional, educational, or
business use. Please contact your local bookseller or the Macmillan Corporate and
Premium Sales Department at (800) 221-7945, extension 5442, or by e-mail at
MacmillanSpecialMarkets@macmillan.com.

First U.S. Edition 2021
Published simultaneously with Profile Books, United Kingdom

Designed by Wilf Dickie

Printed in the United States of America

1 3 5 7 9 10 8 6 4 2

To my beloved father and mother, in the hope that we can find a common language in which to talk about the great adventure of being human.

I'm looking for the face I had
Before the world was made.
W. B. Yeats

Contents

BEING A HUMAN

AUTHOR'S NOTE

Few of us have any idea what sort of creatures we are.

If we don't know what we are, how can we know how we should act? How can we know what will really make us happy; what will make us thrive? This book is my attempt to find out what humans are. It matters urgently to me because, despite what my children tell me, I am a human.

I thought that if I knew where I came from, that might shed some light on what I am.

I can't inhabit all human history. I can't even inhabit my own. So I have tried to inhabit three pivotal times by immersing myself in the sensations, places and ideas that characterised them. It's a prolonged thought experiment and non-thought experiment, set in woods, waves, moorlands, schools, abattoirs, wattle-and-daub huts, hospitals, rivers, cemeteries, caves, farms, kitchens, the bodies of crows, museums, beaches, laboratories, medieval dining halls, Basque eating-houses, fox-hunts, temples, deserted Middle Eastern cities and shamans' caravans.

The first of those times is the early Upper Palaeolithic (from around 35,000–40,000 years ago), when 'behavioural modernity' appeared. This is a confusing label. As we will see, today's humans behave (even if they don't think or feel) in a dramatically different way from Upper Palaeolithic hunter–gatherers. Just what is meant by 'behavioural modernity', and where it evolved, are bitterly contentious, but the arguments don't matter for my purposes.

Hunter–gatherers were – and the few that survive often

are – wanderers, intimately, reverently and often ecstatically connected to lots of land and many species. They lived long and relatively disease-free lives, and there is little evidence of human–human violence. For most, settlement wasn't an option, and even if it had been, it would have been unappealing. Why chew on rusk all your life when you can graze from a vast, succulent and ever-changing buffet?

It was unusual to own much more than a flint knife or a caribou-scrotum pouch. If you knew as much as humans then did about the transience of things, it was ridiculous to assert ownership: the world isn't the sort of place that can be owned, and they (unlike us) thought that humans shouldn't behave in a way inconsistent with the way the world is.

It was a time of leisure. You can't hunt or gather all day and all night. And so, I think, it was a time of reflection, of story, of trying to make sense of things. The earliest human art, on the cave walls of southern Europe, is among the best there has ever been. It is also the most allusive and elusive.

To those who might suggest that this is romantic noble-savagery, for the moment I'll just say that I don't see that the allegation 'Romantic' needs a defence. 'Romantic' isn't a term of abuse. Quite the opposite. Romantics just take more data into account in construing the world than do their opponents.

The second period is the Neolithic, which is conventionally supposed to have started around 10,000–12,000 years ago, and lasted until the dawn of the Bronze Age – say 5,300 years ago. Chronology is contentious, and the transitions between phases vary significantly between regions, and of course there's no clear boundary between the different eras.

Some hunter–gatherers started to settle for some of the year, while continuing to wander for the rest. No doubt they started to manage the land – perhaps by planting trees whose fruit they liked to eat – long before anything like systematic farming happened. But nonetheless the division was eventually very real. Wanderers stopped wandering. Their geographical world became smaller. They no longer had to know and relate to a huge number of species. They could get by – and eventually *had* to get by – by knowing just the cow (a subdued and truncated aurochs) in the field behind the hut, and one species of specialised grass with big seeds. It didn't take long – a few thousand years – before everything, including humans, was subdued and truncated. The relationship with the natural world was changed from one of awe for, and dependence on, everything to control of a few square feet and a few species.

Although the mental attitude of the Neolithic was hubristic control, the reality was very different. Humans began *to be controlled*. They had to stay in their settlements: they had to get in the harvest. Settlement brought politics, hierarchy and man-made law. Lives shortened. Pestilence galloped. Bones were distorted by strenuous grinding and lifting. The enslavers of pigs and reapers of corn were themselves enslaved and scythed down. The cycle of the seasons, which had previously propelled them, now ground them down, and they were tyrannised, not enriched, by the law of supply and demand. Leisure vanished. Hubris gets you in the end: ask any Greek. The grand tales of the Upper Palaeolithic were codified and constricted into the priest-curated stories of Stonehenge. Codification and constriction strangle the mind. Thoughts as well as sheep were corralled. We see the marks of strangulation in the art. Neolithic art is less accomplished, nuanced and evocative than the Upper Palaeolithic art that preceded it. In the Neolithic we started to get boring and miserable.

The final period, in which, despite some spirited resistance, we still are, is the ironically named Enlightenment. The Enlightenment continued and systematised the revolution that had begun in the Neolithic. The divorce proceedings between humans and the natural world, filed in the Neolithic, were completed. The decree nisi was the corpus of Descartes's writing; the decree absolute was signed by Kant. The result was the systematic desoulment of the universe. Until then (and, yes, even in the Abrahamic monotheisms) everything had been pregnant with some sort of soul. Aristotle had insisted on it, Eastern Orthodoxy never doubted it for a moment, St Thomas Aquinas made it canonical for Catholics, the kabbalists catalogued it, and the Sufis danced it.

The Enlightenment abolished souls from the non-human world. The universe was now a machine, governed not by some embodied essence, but by the laws of nature. Laws are a lot less interesting than essences.

Because the Enlightenment was, at the beginning, a revolution in Christian brains, humans were allowed to hang on to their souls for a while. But not for long: soon we were left as machines in a machine. The slogan 'rage against the machine' shows a very precise understanding of what has happened since the seventeenth century.

Darwin might have mitigated some of the disaster. He reminded us that we're part of the natural world – which was the central insight of the Upper Palaeolithic. That, properly handled, could have generated a fitting humility. But, by and large, this part of Darwin's message was transmuted into a cynical and dangerous 'nothing-buttery'. He was (wrongly) heard to say that humans are nothing but 'cogs in the machine': that there is nothing but matter, and so nothing matters. It was a recipe for low self-esteem and wanton environmental destruction. It might be wrong to kill something that has a soul: there's nothing obviously immoral about smashing up a machine.

It was now logical enough (and chimed well with Darwinism's identification of competition as the fuel that powered the world's engine) to see humans as *Homo economicus*. They had long been, in various iterations, *Homo deus*. One of the clearest indicators of behaviourally modern humans in the archaeological record (and certainly the most repercussive and *defining* of indicators) is religion. If there is clear evidence of religious practice in your excavation, you're dealing with behaviourally modern humans.

Now God was gone. There was only matter, and we were only matter. Nature was, like us, red in tooth and claw but could, like a circus lion, be very valuable if treated firmly. The only metric of value in this world was economic. There were no longer intricate, ancient and heart-breakingly beautiful natural communities: there were, instead, *natural resources*. The idea is so ingrained now, even in the discourse of conservationists, that it doesn't grate on us. Why should an ancient grassland be preserved? The answer we hear is that it has a value in dollars.

The best hope for us, since Enlightenment reductionism has metastasised so far through our culture's vital organs, is probably the Enlightenment itself. Scepticism and rigorous empiricism were central to the original Enlightenment manifesto. We see neither in the citadels of the modern Enlightenment – such as actuaries' offices and most biological research labs. But scepticism and empiricism can and must help us to recover enchantment. If we're sufficiently sceptical and empirical about *anything* (whether a star or a baby or a plastic cup), we'll see that it's baffling, mysterious, thrillingly weird and defies all our categories – requiring a poetical and a mathematical and an emotional and a physical response. Scepticism and empiricism, properly deployed, expose the vertiginous wonder of the world – a wonder that demands all our resources, in all our intellectual and sensory and, yes, spiritual modalities, for its exploration.

So this is no anti-Enlightenment tract. Far from it. It's a

plea for the Enlightenment to do thoroughly and honestly the task it set itself in the eighteenth century. It's an attempt to prise the Enlightenment out of the clutches of its self-appointed High Priests – the scientific fundamentalists – and to get it to look fearlessly and without bias at the natural and human worlds. If it does, it will join Niels Bohr (who demonstrated that uncertainty wasn't a failure of science but part of the very weave of the universe), Werner Heisenberg (who knew that scientific objectivity was impossible because all observation is coloured by the relationship between the observer and the observed) and the shamanic painters of the Upper Palaeolithic (who knew, just like Darwin, that the boundary between humans and non-humans was fluid) in a vibrant scientific mysticism. If science addresses itself properly to its subject of real existence rather than to neurotic affirmation of its own presumptions, it will be an epic and mystical calling, for existence is epic and reality is mysterious.

We are materially richer than ever before. We have abolished many material ills. And yet we are ontologically queasy. We feel that we're significant creatures, but have no way of describing that significance. Most of us abjure the crass fundamentalisms – both religious and secular – that give us cheap and easy answers to the question 'Why am I alive?' No Upper Palaeolithic hunter, looking up at the sky, would demean the gods by thinking that they could be constrained within the terse formulae of conservative Protestantism.

We are laughably maladapted to our current lives. We eat in a single breakfast the sugar that an Upper Palaeolithic man might eat in a year, and wonder why we're diabetic, why our coronary arteries sludge up and why we're tense with unexpended energy. We walk in a year what an Upper Palaeolithic hunter would walk in a day, and wonder why our bodies are like putty. We devote to TV brains designed for constant alertness against wolves, and wonder why there's a nagging sense of dissatisfaction. We agree to be led by self-serving sociopaths who wouldn't survive a day in the forest, and wonder why our societies are wretched and our self-esteem low. We, who work best in families and communities of up to 150, elect to live in vast conglomerations, and wonder why we feel alienated. We have guts built for organic berries, organic elk and organic mushrooms, and we wonder why those guts rebel at organophosphates and herbicides. We're homeotherms, and wonder why our whole metabolism goes haywire when we delegate our thermoregulation to buildings. We're wild creatures, designed for constant ecstatic contact with earth, heaven, trees and gods, and wonder why lives built on the premise that we are mere machines, and spent in centrally heated, electronically lit greenhouses, seem sub-optimal. We have brains shaped and expanded, very expensively, for relationality, and wonder why we're unhappy in an economic structure built on the assumption that we're walled islands who do not and should not bleed into one another. We are people who need stories as we need air, and whose only available story is the dreary, demeaning dialectic of the free market.

These last observations about the state of the world are trite. What is not so trite is their connection to the last 40,000 or so years of human history.

This is a travel book. The travel is into the past, in an attempt to discover what humans are: what the *self* is; how the past is connected to what we are now. It's an attempt by one man to *feel* that connection: the story of how I tried to turn myself into a hunter–gatherer, a farmer and an Enlightenment reductionist – all in a desperate search to know what I am, how I should live and what shape consciousness adopts when it is folded into a human body.

I think it was worthwhile. It was certainly fun.

Scholarly books about the past start with facts: I start with my feelings – feelings that occur when I've immersed myself as best I can in an era, or a wood, or an idea, or a river.

After all, they *felt* things in prehistory and the Enlightenment, and we'll be better scholars of those ages if we have a better idea of what those feelings were.

Nothing is ever of *merely* historical interest, and the study of our formative years certainly isn't. These ages haven't passed: they still rule us. I tend to get on best with my explicitly Upper Palaeolithic friends – the ones who don't know where they end and the garden begins – but most of us, at least in the early hours of the morning, have Upper Palaeolithic reflexes. Our tameness and our lust for compartmentalisation, dominion and control are Neolithic, and blight us and everything we touch. But the Neolithic parts of us aren't all bad. From the Neolithic comes our desire to cosset and nurture the earth. Only Neolithic people buy birdbaths and dogs.

The book isn't a manual. You will find no recipes for reindeer fricassee, patterns for goose-skin leggings or instructions about how to carry fire in balls of fungus, bind a flint axe-head to a shaft or raise a standing stone. Nor is it a record of a systematic attempt to re-enact life in other ages. There are plenty of books and web sites that do those things.

I am not an archaeologist or anthropologist, but I have tried to get the facts right (or at least not to get them wrong) and not to misrepresent scholarly consensus where it exists. Some of the leading figures in prehistoric archaeology and in anthropology generously and patiently sat me down and answered my questions and tried to put me right. They are listed and thanked in the Acknowledgements. If they failed to put me right, where that was possible, that's entirely my fault. It needs to be understood, though, that very often there are no 'right' answers in questions relating to human prehistory. There is enormous room for opinion, and very often, I've found, opinions are dictated just as much by the temperament or personal history of the protagonist as they are by what has been dug up. It's the same in most regions of academia, of course, but perhaps it's more visible in pre-historic archaeology.

The conversations in the Enlightenment section are culled from many I've had with many people over many years. You will search the Oxford cloisters in vain for the Professor, the Shakespearean and the Physiologist. Or rather, they are everywhere. Nowhere and everywhere, too, are Steve the Peedo and his fellow slaughterers, Giles the neo-Neolithic Christian farmer and the capitalistic Master of Foxhounds.

At various points throughout the book we will meet two characters from the Upper Palaeolithic: a man whom I call X, and his son. I have been asked if they are real: if I really met them in the wood, and if they really reappeared sub-sequently, commenting wryly but silently, representing the voice of normative, pristine, newly hatched humans, unsul-lied by the compromises of 40,000 years, or whether they are just a device. To which I reply: first, that I'm not sure. And, second: a plague on your dichotomies.

The Upper Palaeolithic section of the book is very much longer than the Neolithic section, which in turn is very much longer than the Enlightenment part. The discrepancies are entirely deliberate. Humans spent far longer being Upper Palaeolithic than they spent being Neolithic, and far longer being Neolithic than they spent being Enlightenment people, and (I argue) the respective contribution of those eras to the kind of animal that we now are is roughly in proportion to the amount of time spent in each phase. Judged by the length of each phase, the Neolithic section is far longer than it should be, and the Enlightenment section is far, far, far too long. If the Upper Palaeolithic started 40,000 years ago (and if we lump the Mesolithic in with it, which seems reasonable for these purposes), and if the Neolithic started 10,000 years ago, and lasted until 5,300 years ago, and the Enlightenment started 300 years ago and continues to today, the Upper Palaeolithic section should be 86 per cent of the book, the Neolithic section around 13 per cent and the Enlightenment section 0.86 per cent. If the Enlightenment section seems like a mere coda, that's because it is. I didn't want to indulge the Enlightenment's fantasy that it is the main theme.

There are historical ages other than these three. Some of them are really quite important. But I deal with 35,000 out of 40,000 years – missing out only 5,000 years, or around 13 per cent of the time that humans have been behaviourally modern. For entirely personal reasons I would love to have dealt with that extraordinary time around the fifth century BCE which saw the birth of the great monotheisms, the articulation of most of the perennial problems of philosophy and the constructions of the foundations of science. But, awestruck though I am by the achievements of that time, I cannot convince myself that it was as formative as the three ages I have chosen. It changed the way we describe ourselves: it didn't change our substance.

The Upper Palaeolithic and Neolithic sections are divided according to the seasons of the year. The Enlightenment section is not. The Enlightenment has no seasons. Seasons happen in the natural world.

I'm aware of the irony of writing a book, in human language, that questions the value of anything said or written in human language. I don't know what to do about that, except to admit that I'm embarrassed.

I speak often about the presence of the dead. Do not hear that as an encouragement to seek to contact the dead. *Do not*: it is desperately dangerous.

I'm often bitterly critical in this book about the way that humans have behaved, but that's because I think they're

glorious. All of them. Our behaviour is often shameful precisely because our real nature is glorious. We fall dreadfully
because we have such a long way to fall. Each life is immensely
significant, and we let ourselves down. We literally *de-mean*
ourselves. I hope that when I criticise I'm doing no more
than recording and lamenting the de-meaning. I hope I don't
come across as angry. I am far more sad than angry – sad at
what might have been. But I am far more excited than sad at
what might yet be.

I don't explore here what is to be done. I am no seer, sage,
shrink or sociologist. But it will involve radical kindness,
waking up, and old stories. All humans are Scheherazades:
we die each morning if we don't have a good story to tell,
and the good ones are all old.

Finally, I shudder at the hubris of trying to say what
humans are. But surely each of us has to try, at least for ourselves, to decide what we are?

UPPER PALAEOLITHIC

WINTER

I first ate a live mammal on a Scottish hill.

A couple of days earlier I'd been standing in a Victorian court in central London wearing a horsehair wig, a stiff wing collar, starched bands and a black gown, arguing about how much a damaged uterus was worth. Then I'd rattled on the sleeper up to Scotland, drinking Chianti, been disgorged at a Highland station, driven in a Land Rover to a big country house, made to shoot at a picture of a charging Russian and released onto the hill in a tweed suit.

For six hours I tramped, scanned and crept. At last I saw a big enough stag and said, 'He's mine.' He was in a hollow just below the crest of the hill, and it was the devil of a job to get up to him. The wind ricocheted off the rocks, and I hoped I was high enough to stop my scent bouncing down to him. I crawled up a burn, with water coming in at my neck and out at my socks, and lay behind a stone for a couple of hours. I couldn't go any further, but if the stag didn't move there was no way I could get a killing shot.

A raven gave me away. It swooped, saw me and croaked. The stag knew something was wrong, stood up, sniffed and gathered his hind legs under him to take off. It was now or never. I raised my head, pushed off the safety catch and squeezed. The bullet took him in the chest.

It was good enough. He coughed and staggered off towards the sea, but he wasn't going far.

We found him jerking in the heather. His brain was electrically dead, and his heart had stopped, but most of the cells in his body were still alive. The stalker, Jimmy, took a knife from his belt, stuck it in the belly and ripped it open. The guts uncoiled and steamed like hot snakes. Jimmy hacked out a piece of liver and handed it to me.

'It's grand just now,' he said.

What was I expected to do? Jimmy cut out another piece and started to chew it, so I chewed on my piece too. One surface was elegantly domed where it had pressed against the diaphragm. It had been pushed down thousands of times a day by a bellows full of salt air from the Outer Isles. Now the whole thing moved like a slug. The end of a tube nipped my tongue and squeezed blood into my mouth.

'Good, eh?' asked Jimmy.

'Great,' I said, trying not to be sick.

There was still blood on my face when I got back to the house. I bathed, changed and went for dinner. It was very fine Burgundy that night, and a beautiful woman sang some Schubert *lieder* at the piano afterwards.

*

The following week I was back in court, wondering aloud about the relevance of an eighteenth-century case to a twentieth-century paediatrician, deafened by the dissonance between the different modes of my life, wondering what sort of thing I was, where I'd come from and what the hell I was going to do about whatever the answers turned out to be.

And then, of course, I did nothing about it for years. The dissonance became an irritating but not particularly intrusive tinnitus. I got on with travelling and killing and reproducing and speechifying and trying to persuade, and, dangerously, I sometimes even persuaded myself. The whirr of the busy ness made it possible to ignore the tinnitus, except in the early hours of the morning, or in the few frightening moments when I was alone. But then, prompted by nothing obvious, it swelled until it filled my head, and I knew I had to do something about it.

*

What I had to do was to start as near the beginning of my story (and your story) as I could: to walk the step, to meet the family, to feel the forces that made me the shape I am. But there are limits. Our start was a mathematical convulsion that became an explosion – an explosion that never happened in time because time hadn't started, and that happened nowhere because space hadn't been invented. You can't start there without going mad.

It would have been just silly to join our story when the family were sponges in the sea off modern Madagascar, or shrews scuttling between the legs of London triceratops. But it wasn't so silly to join it 40,000 years ago, when humans, with bodies and brains as modern as yours and mine (just better), lived in caves and shelters in Derbyshire.

It was cold then. There was shrill, bleak tundra rather than the dense forest that came when the last ice gave up. The men grew beards, and their hair hung low on their shoulders, but their bodies were as hairless as mine, though harder. They wore neatly tailored hide clothes, had roasts on Sundays, loved their children and didn't want to die.

There was one big difference between them and us. Their sense of self wasn't as intrusive and tyrannous as ours. If they had some sort of language (which they did – but more on that later), they didn't befoul every sentence with 'I' and 'me' and 'mine'.

Up near my friend Sarah's Peak District farm there's a wood. I think one of these men lived there with his son, when there were only little, scrubby, occasional trees and tough grass. I daren't call this man by a modern name. I'll call him X. If I can find him and look him in the eye, I'll know what I am.

Perhaps one day I will know his name.

★

Tom and I take a train for 150 miles and 40,000 years. We change in Derby, where we drink tea, play cards and finish off a flint spearhead.

'Very irresponsible,' a highly scented woman had said the week before. 'By all means indulge your own fetish for squalor, and your own perverted ideas about cavemen as Philosopher-Kings, but don't force poor Tom to come along too.'

'Have you seen the forecast?' asked a man with little hot

eyes who believes what he reads in newspapers and plans to hold his wife's wake in an airport hotel. 'Sounds like a case for Social Services.' He was serious, according to his brow and his CV.

Tom is thirteen, and wonders what all the fuss is about. We'd lived in holes before. Now we're going to a wood we know well, and we're going to make a shelter, kill things and stare into fires until Christmas. Then we'll get the train back in time for all the usual things.

His teachers were understanding: 'Interesting time, the Upper Palaeolithic. Try to keep up with your maths, won't you?' His mother wasn't. 'Do you know how behind he is already?'

We've heard all the jokes about mammoths ever cracked. Our faces are tired from forced smiles.

At a tiny country station the taxi that's going to drive us up to the moor is waiting. A plastic dog wobbles on the dashboard.

'Have you got a dog?' I ask the driver.

'No,' he says, and that's that.

We go a silent mile before Tom says: 'Can we stop, please?' So we do, and out Tom hops with a black bin liner, puts a dead fox in it just like I did at his age, gets back in and puts on his seat belt.

'Thank you,' he says. 'And sorry.'

'No problem,' says the driver. 'Just try to keep its bowel contents off my carpets.' He has the professional detachment of a priest in a confessional.

We stop twice more, but for dead rabbits. Their eyes have shrunk back into the sockets and have films across them, as if they're watching something inside themselves; as if grass-eating and mating are dull compared with what's playing now.

The taxi winds up the dale past chip shops, abandoned

mills and standing stones. Festive lights blink around plastic windows. Hot air, stinking of diesel, pumps out round our feet, and the fox's musk rises and surges round the car. Cackling drunks stagger out into the road. The driver swerves around them without comment.

The street lights surrender. The dark is bigger than them. We plunge on through a tunnel bored by the headlights into the night, and when the hill rears up we drive towards the sky. The road flattens onto the moor, or what they call the moor up here: fields of thin grass littered with sheep bones, etched by drystone walls built by men buried in the field corners. There's always wind here. It glances off the walls like a squash ball, so it's always coming at you from all directions at once.

The cab drops us by a gap in one of the walls. We sling our rucksacks and the roadkill onto the verge.

'Have fun, won't you?' says the driver as I pay him. He's not smiling.

I can see Sarah's TV flickering through the windows of her farm, just down the lane. Sheep freeze in the torch beam: clouds of wet wool, green with algae. Our breath hangs round our chins.

We bang on the door and, when we get no reply, on the window. Sarah must be in the pub three miles off down the dale. It's curry night, and a band from Sheffield is playing bluegrass. On the screen a narcissistic psychopath is threatening to beat up a little country. A cookbook is open on the sofa by the TV, and a cat is rubbing itself against a vat of kombucha. The oranges in the fruit basket are from Israel, and the light from last week's wind. In the kitchen a grouse is rotting until it's edible. We try the door, thinking that we'll raid the fridge and perhaps sit by the fire. It's locked.

We've heard that a big black dog of a storm is about to bound in from the north, and now we scurry to get under cover before he arrives. Through the gate; down the hill; avoid the mine shaft to the left; past the hare's tree; remember to bend low or the thorns will get you in the eye; have a piss before you dive into the wood; note the pheasant exploding from the rowan (we'll be back for you, my lad); don't worry about your coat on the spikes, Tom; let's just get our heads down. Under the low long branch of an old hawthorn, stretching out beside a collapsing wall. Out of here, you sheep: this is our place now. Get out and take your ticks with you.

The storm dog comes. He doesn't even growl first. He's suddenly there under the tree, snarling and snapping; all hair and spittle. We'd planned to tie a tarpaulin to a tree to make a tent until we could make a more authentic shelter, but he'd rip it straight off. So we curl up on the ground as near to the wall as we can, wrap ourselves in the tarp and let him do his worst.

It's not so bad, though it's pointless trying to sleep while he's in the wood. He roots around for a few hours, trying to get to us. He slaps us around with his paws for a bit and then, frustrated, cocks his leg over us and moves on to see what he can find in Nottingham.

When he leaves, the wood sighs, shakes itself and breathes again. A damp owl kills. Badgers lumber through the brush, sucking up worms like spaghetti. A sheep coughs. There are no stars. Cold crawls from the earth and through our clothes. We think about fire and tea and wine. Sleep creeps into us with the cold. We are part of the mud.

When I wake, the fox is my pillow. It is blue and white and shining out there, we're in a wood on top of the world, and we can begin.

★

What we are beginning is *ourselves*. By which I mean ourselves as modern humans.

Here's the ruling theory of mainstream anthropology. Human evolution began in Africa. There were several proto- types, some of which co-existed. They were all brutally road-tested by natural selection. Around 200,000 years ago *we* appear in the fossil record: beings, that is, anatomically and physiologically more or less identical to us. Their brains were the same shape as ours, but perhaps slightly bigger. You need a big brain for relationship, which is costly, demanding and very rewarding. They did relationship better than we do, and so needed powerful neurological hardware. They stalked around the veldt on two long, strong legs, gazing with their forward-facing binocular eyes at horizons that would have been hidden to their non-bipedal ancestors by the long grass; looking down at the world literally and, eventually, figura- tively; seeing the earth at their feet; blessed and cursed by a perspective that nothing had previously had; their noses, hoisted out of the dust, subservient to their sight; their clever hands with opposing thumbs freed to make tools, signal, bludgeon and caress, but never again to soak up sensation from the ground.

But anatomy and physiology aren't everything. For 150,000 years these humans didn't behave much like us. They weren't, to use the phrase beloved and hated by archaeolo- gists, 'behaviourally modern'. Probably they didn't adorn their bodies, bury their dead with grave goods, make bladed or bone tools, fish, move resources significant distances, co- operate with anyone to whom they weren't closely related, and probably they weren't organised enough to kill large animals.

Then something big happened. The speed with which it happened, and the amount that happened in Africa, are con- tested. That it did happen is not.

Go to a good museum, and find the gallery dealing with early humans. There will be lots of flint in it. Start at the beginning, and walk chronologically towards today. Look carefully at the artefacts. The first few minutes of your walk will be really boring. You'll be looking at dowdy things: lumpen, undifferentiated tools and pictures of hairy men roasting carrion. Everything is directed relentlessly towards the material. Everything behind the glass says that humans are just lumps of meat, bone and gristle.

Then, if you're in a really good museum, you'll turn a corner, the labels will say 'Upper Palaeolithic', and your heart will start pounding. For there, from around 40,000 years ago, you'll see the family. You'll recognise it by a massive explosion of symbolism. Bone and stone are made to stand for wolves, bears and humans, and metaphor must have been born at the same time. There was a tsunami of possibility. A bone could be a wolf while still being a bone. If that was possible, was anything impossible? A world that had been merely chemical had become alchemical. Just because something was invisible according to the laws of optics and visual physiology didn't mean that it didn't exist.

Time started to behave differently. Now it didn't seem to be the normal medium in which humans swim, and no doubt there was a revolt, previously unthinkable, against the tyranny of time, or at least against the notion that one moment just plodded in front of another. The dead continued. Human dead were anointed with ochre and sent on their journey with food, weapons and objects of merely sentimental or aesthetic significance. The animal dead were appeased. As a bone could also be a wolf, so the dead could be both on a journey and present at the campfire to comfort, advise, rebuke and taunt.

The world was hugely more complex and resonant than it had been.

Tom and I hope we're in that world now, or can find a way there. X is here somewhere, waiting to help.

This new complexity demanded and gave more. It takes a prism to show that white light is anything but white: that it's composed of many colours. This was the new prismatic age. What in the old days had been one job (hacking up a dead bear, let's say) was now many: skin the bear, cure the skin, dissect out the tendons and make strings, turn the thigh bone into a hyena, make the dead bear safe, and ideally friendly. So we see, for the first time, elaborate tool kits containing specific tools for specific tasks. There was a new precision in action and thought. Blades were used to cut precisely along a pre-planned route through joint and organ where, before, a blunt axe had crushed and splintered.

To pre-plan the passage of a flint blade through a bear's belly, and to discard other possible routes, means that possibilities have been created and evaluated on some virtual drawing board. There was, in other words, abstraction: a move away from the concrete world of stone and bone to another arena of action – a place, like the realm of the dead, which could not be seen with physical eyes, but which was real nonetheless, and whose results could be seen when stone was hammered or bone trimmed.

Abstraction conferred massive advantages. The various strategies for killing a bear could be examined in the safety of your head, rather than tried out, dangerously, against a real bear in a cave – when you'd have only one chance to get it right. It could not have been done without the idea of an 'I'. An 'I' had to be the main actor in the imagined drama: An 'I' had to throw the spear and avoid the paws. An 'I' involves looking down at oneself and describing one's self to oneself.

There's a word that means, literally, standing outside oneself: it is *ec-stasy*. The etymology's curious. Do you have to get outside yourself to have ultimately pleasurable feelings? Well, that's my experience. Selfish bastards are miserable bastards. Do you have to have distance for a proper view of yourself? Yes again. But a proper view of myself certainly doesn't produce pleasure. Perhaps the Greek self-viewers who coined the word *ecstasy* were nicer than I am.

This ec-stasy – the self-seeing – of the Upper Palaeolithic is crystallised in the museums' bone figurines. Human faces appear for the first time. They are the most eloquent art ever, and they shout out: 'This is me' or 'This is you, and I'm different from you.'

What follows from this? Above all: *story*. You and I are actors. Actors don't just stand around with their hands in their pockets. They can't. They act compulsively, and their acts are joined together, creating story. Little parochial stories give rise to bigger stories. If you see what happens to other humans when they're taken by a rock fall, lion's teeth or a speeding Merc, only years of arduous conditioning can stop you telling a story that makes sense of you and your own annihilation.

The 'I' birthed the 'you'. The way was open for the human variety of Theory of Mind, and hence for our kinds of love, empathy and acquisitiveness. The ancient desire to kill was recast, and this time it had moral colour. There was a nagging hint, which turned into a deafening shout, that some sorts of killing might be wrong. A sense of self spawns all laws, all ethics, all sadism, all love and all war.

As soon as there was an 'I', existence stopped being merely a chain of events. It remained like that for another 45,000 years or so, and then, as we'll see, we were told that there were no stories but only events: that *we* were only events – chemical events and their corollaries. A few people even believed it. Human consciousness was first demonstrated by symbolism: by things that signified other things. Now, we're told, nothing signifies anything. Nothing is significant.

The 'I' revolution may not have come just once to humans, in one great surge of self-creation and self-knowing. It may have erupted many times, around many different campfires, and over many thousands of years. But however, wherever and whenever it happened, it made *you*.

★

It's strange that X and his son are here at all. This was the very edge of the world and the edge of the ice, a place of teeth and skirling snow and screaming wind. They would need a good reason to be away from their home in France. Perhaps they told their family that they were going on a mammoth hunting trip, but if this was to be at all credible they would have had to come with a number of others, and I've seen no sign of them. Back home, X would be in a hunting or foraging group of about fifteen, part of a clan of around a hundred and fifty, and connected to a network (with whom he shared language) of about five hundred. He'd see most of the five hundred only occasionally. They would be distant points of flickering light, or a column of smoke, or a festering flint spear wound in the side of a caribou or (if hunger or wolves had come) circling kites and ravens. They had different scar patterns hacked into their faces, different ways of furling cloaks, squatting to defecate, marking trails, filling sausages, copulating and thinking about the constellations. They rarely fought with X's clan. Why would they? There was plenty to go round, and fighting hurts. But sometimes their different smell marched into X's valley (though they themselves never came), and then X gripped his knife and shook his son awake.

I suspect that X and his son were really here because X had begun to feel a gentle itch in his mind: hear a quiet whisper. The itch and the whisper seemed significant, but they couldn't be examined when family obligations pressed hard: when children needed to be taught, wives obeyed and grandparents fed.

At last X had to be alone to scratch the itch and hear the voice. Being alone meant being by himself, but since he was only himself with his son, they went off together into the cold.

Sitting by the fire, he could hear the voice. 'Me, me, me,' it said, and as he listened it got louder until it drowned out the roar of the wind and the hiss of the wood and split open his head and his world.

★

I'm standing outside myself this morning, just as my ancestors stood so fatefully outside themselves. I'm used to it, and it's a burden. For them, that first time, it remade the universe.

I see a battered, fearful, proud, tall, bearded animal in an old tweed jacket, with a set of Greek worry beads and the complete poems of Thomas Hardy in his pocket, living not in a Derbyshire wood in December, but in the past and the future: in virtual realities conjured by his autocratic brain. He likes to think that he uses the abstractions as tools, but really he's their tool.

But Tom's in the wood, and he's here *now*. He climbs a tree and falls out of it, and has real honest pain in his elbow in real time. He digs for voles as dogs dig – scrabbling with his hands and sending the soil through his legs. He sucks soil from his fingers and says it tastes of moles. He laughs as the sun is splintered by the arc of his piss, tries to hypnotise a robin and, before I can stop him, nearly spears a sheep. He laps from the pond, strokes beetles, keeps an earwig in a bottle, gives companionable names to birds and trees, and turns a stone in his hand for an hour, wetting it with his spit to release the smell of Carboniferous ferns.

He has the great gift of dyslexia. I don't. He's a linguistic cripple, and so a sensory and ontological athlete. When he goes into a wood, he sees a tree. When I go into a wood, photons stream from a tree and hit my retina. So far so good. But then data flow along my optic nerve into my brain, and the trouble starts. For I translate those data almost immediately into things that have nothing whatever to do with that tree: into remembered fragments of poems about trees; into generic physiological facts about trees. When I say, 'That's a tree!' I'm lying or deluded. It's no such thing. I've never seen a tree. In fact I've never (at least for decades) seen anything at all. I bet you haven't either. I once met an adult who did see trees, and it excited and scared me so much that I bolted straight for the airport, leaving my bags and my girlfriend at a mountain monastery. I'm locked into my own head. I'm wholly self-referential and hence self-reverential. This is dangerous and dull. I'd love to see a tree. From what I've heard, they're much more interesting, colourful and charismatic than my thoughts about them.

Tom has seen lots of trees. I hope he can help me to see one too.

X had a kernel of words before he left the clan and came here. They were blunt, curt, useful words: rough flint hand-axes rather than knives or needles. When he was alone with his son in the winter, staring at the ice and his hands, the words lay unused in his head. As they lay, they acquired patina. They were colonised by the slow-growing lichen of association and began to be complex and to resonate at the frequencies of frogs' throats and trembling grass. Then they bred, and when, in the spring, the snow cleared and X went back to the dripping gritstone hole where his family waited, the words spilled out of him and infected the whole clan.

★

'Is this about any more than just messing around in a wood, Dad?' asks Tom.

It's a very good question.

He goes on: 'It's just camping, really, isn't it? Without toilets.'

'Or tents. Or food,' I volunteer, trying to convince myself.

The truth is that we *are* just messing around. It's better than not messing around, but we're not really living as hunter–gatherers. Being cold and squalid is part of my normal life. The main difference is that hunter–gatherers had to live like this. We don't. We have plenty of alternatives, though, in search of sensation, we might choose not to use them. The knowledge that there are beans and crisps in the village shop, and a roof and a bed a few hours down the line in Oxford, means that we're bogus.

Yet in some ways we're subject to the caprice of the wild world just as reindeer-hunters were and are. An electrical storm in an old friend's heart saw him off a few months ago. He wasn't given any choice. That's not so different from a lightning strike. The uncontrolled multiplication of another friend's gut cells took him to the graveyard gate, and he returned without his colon, his faith in the goodness of the gods, or his hair. That's not so different from a near-miss with the local wolf pack. And the uncontrolled multiplication of neuroses in my own head – neuroses that stamp 'Conditional' on every future diary entry: that's not so different from the caveman's knowledge that he's stuck on his howling plain whatever it might do to him: that the hand has been dealt, and even the canniest player can't make anything of a bad hand.

I know something of what it's like to live hand to mouth and day to day. I've wandered in deserts where, if I didn't reach the next water hole, or if it was dry, I'd die quite nastily.

I've never had a regular pay cheque. I've sailed and swum across seas that wanted to eat me. Once you've sniffed contingency you smell it always. It's here in the wood.

So is X. Last night he was sniffing at the bottom of a wall, where the sour belch of a roe deer was trapped. I can never hear him move. He must have soft caribou-hide shoes, tanned with his wife's urine, and they make no noise on Sarah's field. He must tiptoe around the sticks in the wood. Last night, though, the boy stumbled, and cursed as he fell.

★

The idea is to be in at the start of human behavioural modernity. To feel the first rush of Self as subjectivity is injected, to watch the first flickerings of consciousness and the first detonation of story, and to be buried in the avalanche of exponentially proliferating possibilities.

That's a big ask for a morbidly over-educated man, a boy, a catapult and a bag of Cornish pasties.

To do it properly we'll have to become unconscious: to wipe our hard drives and hope that we can be re-booted – no, booted.

Then I'll try to describe the brave new world that we've known with our virginal eyes, noses, ears and psyches.

All this is tricky. For along with the pasties and the catapult, we are taking *ourselves* with us into the wood – cores of consciousness, encrusted with memories and traits: selves interpreting themselves to themselves with a lot of deeply embedded language.

★

A few years ago, leaping (I thought dashingly) onto a stage to give a lecture, I slipped and dislocated my shoulder. It hurt.

They drove me off to hospital and tried to put it back. To do this they gave me a mixture of nitrous oxide and air – the 'gas and air' well known to labouring women. It prised apart two bits of me. One part of me rose out of my body and looked down at it as it sweated and screamed, and at the nurse trying to pull the bone back into joint. 'I' could see the deformed shoulder, the freckles on the top of my bald head and the nurse's neat parting. The body was in pain. The real 'I' – the one doing the watching – was not. It knew that the body was hurting, and regretted it and wanted the pain to stop – but rather distantly, as one regrets a cyclone in Mozambique. The ethereal 'I' had, so far as I can tell, all 'my' distinguishing attributes. It was embarrassed by the body's moans, and it felt sorry for the nurse who, just at the end of his shift, had been lumbered with a case like this. It missed its family, and wondered if its daughter's cold was better, and if its mother was going to get any sleep that night. Though it had no body, it still had appetites, and was looking forward to walking up a hill and eating porridge. Then the nurse wrenched too hard, the body shrieked and slumped forward, the tube fell out of its mouth, and I and my body slowly coincided again.

It's been similar, but less dramatic, on medical opiates. I've had industrial doses of them for bones pulverised on cliffs and in the sea. They didn't make the 'I' rise like smoke: it remained indoors. But its concerns weren't the same as the body's. It too didn't much 'mind' (what a big word that is) the shrieking of the neurones, though it could hear them all right. Morphine stops you minding. So do the endogenous opiates that we squirt into our own blood when we drop a brick on our foot. If it's properly trained, Mind itself, whatever that is, can stop you minding.

What does that interesting evening in an Oxford hospital have to do with the beginnings of human consciousness? Just one thing: it suggests that whatever *I* am can move in ways incomprehensible to us. If I can hover over a nurse's centre parting in a hospital, perhaps I can pass through walls, see through them, survive cremation, keep up my trousers with Orion's belt or, more prosaically, take control of the body of a moose.

Think about the cave paintings of Upper Palaeolithic Europe. Most are of animals, and many are dazzlingly accomplished. The artists knew how aurochs stood, how deer throw back their heads in panic and how the intestines of disembowelled bison uncoil. The artists were superb naturalists, but the walls are not simple bestiaries. Galloping among the naturalistic animals are monstrosities: therianthropes (part-human, part-animal hybrids) and chimeras assembled from the parts of different animals. The artists use the rock's natural features to give their animals life. A bulge in the rock becomes a head or a muscle. You can't get away from the feeling that the animal is pushing through into the cave from a world on the other side of the wall.

The animals are never shown running on the ground. Indeed, they never have any real spatial context other than the cave wall itself. You never see them against a mountain or a tree, or fording a river. They seem to float.

There are other things on the walls: banks of wavy and zigzag lines, chequerboard patterns, ladders, webs, honeycombs and dots – often superimposed on the laboriously executed animal figures. Are they simple vandalistic graffiti painted by Upper Palaeolithic yobs? If they are, the yobs got bolder or busier as the Upper Palaeolithic wore on. By the time we get to the Magdalenian era (about 12,000 to 17,000 years ago), the patterns are everywhere.

Paintings from this period are common in Africa too – indeed they were done there until the nineteenth century CE – but they tend to be found on open, sheltered rock faces rather than in the depths of caves. In Africa human figures are much more common: there are almost as many as there are animal figures. Human figures are very rare in Europe.

The African humanoid figures, the work of master artists, are often bizarrely elongated. They often bend at the waist or hold their arms behind their back, and some have obviously erect penises. Sometimes they are pierced with spears or arrows; sometimes something is pouring out of their nostrils. Sometimes they are pulling animals on ropes. Therianthropes and chimeras are ubiquitous, and so are the geometric patterns.

What's going on? There are four main ideas. The first is 'Don't know', and is always to be taken very seriously. The second is 'Art for art's sake', and is palpable nonsense. Many of the European paintings are in places that are immensely difficult and sometimes dangerous to reach. The best-known example is the 'Shaft of the Dead Man' in Lascaux in south-west France. You squeeze through a tiny crack in the rock, and then have to be lowered five metres down a sheer drop to a ledge. (Try doing that in the dark, or carrying a taper soaked with fat from an aurochs's kidneys.) Only then can you see the bird-headed man with four digits on his hand, about to be gored by a dying bison. It's no place for an art gallery, and it wasn't one. The 'pure art' idea can't explain, either, the existence or nature of the geometric patterns, let alone their juxtaposition with the images. It is contradicted, too, by the fact that many of the images are painted directly on top of earlier images, though there's plenty of unused wall space.

The third theory, 'hunting magic', used to be popular, but explains the geometric patterns no better than the 'pure art' hypothesis. Nor can it account for the dearth of missiles. If the idea of a painting was to put some sort of spell on a hunted animal so as to guarantee success in the hunt, you'd expect to see many of the animals bristling with spears or arrows. But very few do – only 3 to 4 per cent. We also have a fair idea of the important prey species, and they're not the ones most commonly seen on the cave walls.

The South African archaeologist David Lewis-Williams probably stumbled on part of the true story. His is the fourth of the theories. He, like us, wondered what to make of the rock art of southern Africa. Then he read interviews with San bushmen recorded in the 1870s, and it all fell into place. The paintings and the engravings, said the bushmen, were made by people 'full of supernatural power' – shamans. The shamans went on their spirit journeys after arduous food-less and drinkless trance dances that might last twenty-four hours. The small blood vessels inside their noses sometimes became fragile from the dehydration and burst – hence the nose bleeds shown in the images. Shamans adopted the bent-over posture in the trance-dance. The therianthropes depicted the shamans just as they assumed or abandoned the animal forms necessary for travel in the spirit world, and many of the other images showed what they had seen when they were there. The rock art images were *travel books*.

The geometric patterns may be the ubiquitous 'entoptic' phenomena associated with many altered states of consciousness. They're what you see when you enter another state of consciousness, with or without the help of mind-altering substances. There's a fair chance we'll all see them when we're dying. Such altered states of consciousness are commonly associated with a feeling that the normal boundaries of the body have been changed – hence the elongated figures. Haven't you ever felt strangely big or small as you've entered or left a dream? And the erections? Penises stood to attention as, in the shamanic trance, ecstatic consummation approached. To enter a woman, say the shamans, is to enter another world, and the unsubtle male body doesn't distinguish neatly between entering a vagina and occupying the body of a wildebeest.

Pierced figures in many parts of the world show shamanic business. Mircea Eliade and Joan Halifax compiled gruesome catalogues of the ordeals endured by shamans – particularly at their initiation. To become a shaman is no easy matter. It doesn't happen in a couple of hours in a tent at a summer festival, however loud the drumming, however strong the cider and whatever the proportion of organic hemp in the average trousers. In the initiatory trance voyage to the spirit world, the apprentice shaman typically experiences torture, dismemberment and death. Her shattered body is then rebuilt and reborn into her terrestrial body. She will never be the same again. Because the other world is her new birth-place, she has a right to be in the abode of the spirits. She has dual citizenship and so can broker deals for her earthbound but spirit-affected clients. She can bring things back from the spirit world – for instance, those animals on ropes.

The very first real evidence of human symbolising and religion and the other things that shout out '*I am*', comes at exactly the same time as evidence of these shamanic jour-neys. This suggests, although of course it can't prove, that the journeys were themselves the *cause* of consciousness. This isn't the wild theorising of self-justifying acid-heads: it has a respectable place in archaeological libraries. Those voices in X's head are telling him to go travelling. To find himself or create himself he'll have to look back at his body from the far side of the cave wall, possibly with antlers sprouting from his temples. To be in his own head he'll have to get off his head.

This seems plausible to me, shivering in our winter wood. Travel broadens the mind: spirit travel could have transformed it. Or created it. Or freed it from the carapace of matter in which it was caged. Perhaps it's a question of getting far enough away from yourself to see yourself. With the gas and air I got about six feet from my body – and that was enough to convince me that my 'I' had a rather different structure from what I'd thought: that it was composed of parts that, in life, health and the absence of drugs, are so intimately entangled that we see them as the same, but which can be levered apart. If six feet of distance can do that, what effect might a view of oneself from a completely different world have on a creature that hadn't previously had the idea of an 'I', and had never gone further than a five-day march after the caribou? If you fly over your house in a plane, you look down, point and say of a barely recognisable box: 'Look! That's my house!' Might the view from another world not produce the fascinated exclamation: 'Look! That's me!'?

★

We can't sleep tonight. 'Are you awake?' I say to Tom every few minutes, for no good reason. 'Yes,' he replies, always without comment.

The dark on my side of the shelter swells and thrums and pushes against us. The trees moan with the pressure: they are not part of this dark but are caught up in it like us. Usually crows are silent at night, but there's one speaking from the direction of a tall birch by the mine workings: speaking with every breath, like a demon's metronome. After a few hours it suddenly stops, as if something has clutched its throat, and the dark stops pumping. It needed the crow's breath to keep it going.

The day comes slowly and grudgingly. We can hear people heading to Manchester to boost the GDP.

★

Even now, in midwinter, there are rose hips, old hawthorn berries, gusts of fieldfares and redwings on the high moor, and ravens and buzzards hunt in the valley at our eye-level. The ravens sound hollow. They must have taken the eyes from the sheep in the field above our camp. A rabbit screeches like tyres on a tight turn. The stoat will drain it from its jugular vein and it will deflate.

We gather what we can for the roof of our shelter. There's not much – just some old bracken and some clumps of shitty wool cut from the sheep last summer to stop the flies getting to them. It's not authentic (we should really kill a deer or two and stitch their skins together), but we're using the tarp as a roof. No right-thinking Upper Palaeolithic man would turn up his nose at it.

Tom's the purist, and he's offended. He'll use only flint for cutting, and he'd rather go hungry than have the pasties. He made his stone tools himself, pounding away in a Norfolk garden, and I think part of his purism comes from the delighted astonishment of any child at seeing that something they do can cause the world to open up before them. He's a fundamentalist, drunk and fanatical from the heady draught of agency.

The fox has opened up before him. He pressed too hard and its colon flopped out. Now he's more careful, lifting up the skin before he slices down into it, taking the cut right up to the throat and well down into the legs. There's a bruise on its chest where the car hit it, and the blood under the skin is like raspberry jam. Tom hangs the fox on a tree, with a stick under one of the leg tendons, and rips off the rest of the skin. Some of it crackles: some rolls off like waves of cream. He turns the legs and the head inside out, and we see the wood through the fox's eye holes. X would have pulled it onto his own head like a jumper and gone to sleep watching the moon through those holes.

Tom wipes the flint knife on the ground and stands back to look at the carcass. Even now it looks more alive than the fittest domestic dog. 'What a machine!' says Tom, admiringly. No hunter–gatherer would have said that.

He moves on to the rabbits. By the second one he's getting good. He knows just where to cut to make the joints burst open. The joint surfaces glisten as the eyes once did. When the skin is off the head, and the eyes have no eyelids, it's hard to feel sentimental about the animal.

I help Tom with one of the rabbits. It feels cleaner to cut with a blade you've made yourself. You're taking responsibility for the cuts, and however bad something is, it's better if you acknowledge that it's your doing. It is cleaner to kill with a blade than a gun, because you can't deceitfully plead the mitigation of distance. You look at the animal, and decide to kill it, and drive in the blade, and it dies there because of you. If you pull a trigger, you can tell yourself that the animal died because of a lot of engineering. You've got co-defendants standing with you in the dock, diluting your own culpability: gunsmiths and gun dealers, ammunition manufacturers and the police for licensing you to kill. Even the possibility that you might have missed helps you to wriggle out of blame. But if you drive a blade in, *you* drive a blade in, and that's it.

We light a fire of hawthorn sticks and cook the rabbits for breakfast. They're thin, there's no fat, and so we scorch them and chew charcoal. There's one each. We don't need to be told not to eat the fox. I'm not sure why that is. But if X had killed a hunter, his life would have been forfeit.

They may have eaten rabbits in Upper Palaeolithic Derbyshire, but the last Ice Age wiped them out, and the ones we have now are descendants of animals introduced by the Romans.

'They do taste pretty old,' says Tom.

They are very old. Like everything else they're made of

powdered stars. The past is everywhere: nestling in our genes, our proteins, our bones and our algorithms. We breathe it. In the breath I took to burn the fuel to write that last sentence I inhaled some of X's last breath. When I bend down to smell the mud, I get an instantaneous transect through the last 500 million years. All those years, piled onto and squeezed into one another, are delivered in one parcel to my nose. I inhabit millions of telescoped years at once. If I could disentangle the years, I'd get sulphur from the birth of a Cenozoic atoll, the halitosis of a pterodactyl, the athlete's foot of a Roman legionary, Malaysian rubber from the sole of an Australian caver's boot, last month's special-in-a-basket from the pub and a blackbird's panic from an hour ago.

★

Sometimes I think I see X. Sometimes there's a figure standing by the barn. I only ever see him at the edge of my vision, and when I spin round he's jumped into the stone. Sometimes there's a smaller figure with him, who must be his son. He stays for a moment after I've turned, as if he wants to be friends. We came here to be helped by them, but it seems the boy wants some sort of help from us.

★

The Upper Palaeolithic humans here oscillated up and down, and up and down Europe, just behind or just ahead of the ice sheets, chasing spring. When the ice melted, they moved into the relieved land. When the land froze, they went to wherever grass and flowers still grew. A few, scorning the ones who had gone back to the fleshpots of the south, probably stayed on the ice sheets, their faces blistered and their eyes aching, living on pride and on biltong that broke their teeth, trading comfort and a full belly for a sense of superiority and the greater ease of tracking in the snow. I know their sort.

Sometimes this dale was under a glacier. Sometimes it wasn't. But whether it was or it wasn't, our shelter would have made a sensible winter base for people who were becoming like us – though it would have been better to have had a cave, or at least a rock wall at our back. Their movements were determined by the animals, and the path up the bottom of the dale, paced now on Bank Holidays by giggling herds of hikers with fluorescent bum bags, was beaten out by the hoofs of migrating caribou. If the wind was up their noses or their arses, the scent of a human crouched on our groundsheet wouldn't reach them, but a flint-tipped spear just might. If the wind was to one side, it was trickier. Even if the wind was in the hunter's face, it could bounce off trees or off the other side of the valley like balls in a pinball machine. Scent gets stirred around in valleys. I've spent ages here trying to map its swirls. You only get one throw or shot in a valley. Miss it, and you'll go hungry, and you'll have to start walking behind the herd.

That's what they did. The caribou had itineraries, and those itineraries became the hunters' too. The caribou might have gone hundreds of miles in a season, the hunters snapping at their heels. When the caribou rested, so did the hunters. When the caribou moved on, bags were packed and tents struck. It was like a marriage – but a dependent, important one, marked with more mutual respect than most marriages I know.

The hunters' relationship wasn't only with the caribou: it was with each piece of ground where they and the caribou placed their feet or slept, with every noseful of air, throbbing with data and with all the living things with whom they were bound in a solemn, shivering contract of crushing and joyous obligation. All living things meant *all things*. For as soon as the hunters found that they themselves had souls, they found that everything else did too. The discovery determined the whole cast of human thought until about four hundred years ago. It created a big and repercussive problem, because humans had to eat, and since everything (including plants) had a soul, that meant that every mouthful was of some ensouled creature. The problem was solved, or at least mitigated, by rules of etiquette, propitiation, supplication and apology that formed the backbone of the moral and – eventually – the religious life. We don't see the problem any more, and so think the rules redundant.

It's easy to romanticise the lives of hunter–gatherers (as I've just done), but impossible to overestimate the significance of the *nature* of their world: instinct with agency and motive; heavy with moral weight and consequent obligation; coruscating with possibility; and housed, from the earliest times, within an overarching story.

'Tell me a story, Dad,' asks Tom.

I don't have any.

⁂

There are no caribou in Derbyshire today. Behaviourally modern humans ate them all and moved on to sheep and the tiny, meek aurochs that were later called cows. Once I spent a summer following the routes I thought the caribou must have taken through the Peak District: along the valley bottoms, over the shoulders and watersheds; through heather, bracken and football pitches; wading through rivers, winding up scree slopes, trudging along roads, kicking bottles, cursing cars and picking up smashed butterflies. It didn't teach me much about caribou, but it taught me how to mourn for lost things – which came in useful a bit later.

With no caribou to follow, and with the roe deer, in the

cold, staying too near the village's kitchen gardens, we follow rabbits, magpies and robins instead. They too have circuits.

There's no such thing as a rabbit, a magpie or a robin. There are only individuals.

There's a big buck rabbit with a rheumy eye and a scar on his nose, who doesn't like getting up in the mornings, and who looks arrogantly at the young things as if high spirits were irresponsible. I watch him swagger out at noon, sniff the heavy air and go back to bed, disgusted at the universe. I lie for five hours in the old nettles opposite his bank until my neck is an established part of the woodlouse landscape. The cantankerous old man comes out at five o'clock, hops into the dark and at eleven prompt, when I hear scrabbling, he looks back at me, outraged by the torch, ostentatiously eats some of his own dung to show how relaxed he is and goes back down the burrow.

There's a magpie with a white patch on her tail, one of an old and violent family from a stockade at the top of the highest blackthorn in the wood. She's gentler than her siblings, and ashamed of them. She looks on when a stricken sheep is being outraged, and her cackle is like the clacking of soft stones. She's up before the others, watching us, more interested in us than in the scraps of food littering the ground.

She's a bird, so she has two sides. Visual information coming in from her left eye isn't integrated with data streaming in from the right. Her right brain could read Dante as her left brain watched the rugby. She moves her head – tick-tock, tick-tock – so that both parts of her can take us in. It's as if all of her – for despite her divided brain, there clearly is a *her*, with appetites, preferences, plans and fears – wants to know us. This is a great comfort. The other magpies don't tick-tock when they're with us. Their left side or their right side is sufficient.

She starts and ends the day at our camp. When we stumble

out from under the tarp, squat in the nettles, chew a rabbit bone and brush our teeth with the end of a stick, she's on the wall beside us, commenting. When the hill sucks in the last of the weak sun, she's on the wall again, tick-tocking, seeing us safely around the fire before she's off. But otherwise she's strenuously independent, with an inflexible routine. First is the field barn – the 'spooky barn', where, when the whole family's here, I tell ghost stories to the children. Sometimes there's a field mouse or shrew there, left by one of the feral cats. She picks at it like a vegetarian embarrassed by the lack of vegetarian options at a party, before holding her nose and gulping it down with a gag. Then she goes to the field at the top of the village, where cows die every year from lead poisoning, to rummage through the patches of bare earth where the badgers have raked for worms and slugs; then up the dale after berries and walkers' sandwiches; then to the tall trees on the ridge, for a sway and some perspective; then to a tree in the middle of our wood, despised by the other magpies, for a nap and the hope of a moribund squirrel; then to the road, where there might be a flat badger and there will be plenty of pulverised pheasants; and then back to us, for tick-tocks and goodnights.

There's a robin. He's a battered thing, who has come off worst in the battle to bequeath his DNA to posterity. He's lost an eye, and I wonder if that means that the right side of his world doesn't exist for him.

He's much more local. He doesn't go out of the wood. Sometimes in the morning he's perched just a few feet away from the magpie, looking at us as hard as he can, twitching up and down on his thin dinosaur legs and then being as still as only something viciously tensed can be. When the magpie leaves, he stays for a while and relaxes a little. Then resolve creeps visibly over him, and he nods and flickers off to look for things to kill. He's easy to follow. He almost beckons, and stops to make it easy for us to keep up. His journey round the wood and back to our fire describes an almost perfect pentagram.

These animals teach us about the wood and the dale, as caribou taught our Upper Palaeolithic forebears about the couple of hundred square miles of central Derbyshire around here, and as my children teach me about the whole world.

But my main teacher is a hare. Probably there were no hares here in the Upper Palaeolithic. No traces of them from then or before have been found in Britain. They may have arrived with the Romans. But they knew the Upper Palaeolithic world of Europe south of here: they looked on with their brown and yellow eyes from a clearing in a German wood as consciousness descended like a cloud or seeped up from the ground. And although the first humans of Derbyshire may not have known them, I feel when I look into the eyes of a hare what those humans must have felt when they looked into the eyes of a caribou. I feel, in other words, hunger (I want to eat it) and fear (because it is a big thing to kill a hare or a caribou).

We've had nothing but rotting pasties, roadkill and boiled rose hips for several days now. I feel hollow and drawn, though when I see myself in puddles I'm jowly, wobbly and pretty contemporary. Tom's doing better. He's free of many of my presumptions, and one of those is the presumption that we should eat at particular times. His present slightness needs little maintenance, but he has few reserves, and when, suddenly, he says he's hungry, it means something. Now, in the winter, it means we have to kill.

This hare, like the rabbit and the birds, has a daily and nightly round. We first saw her (I hold to the beaglers' convention of calling all hares 'her') last summer in a depression caused by a collapsed mine shaft in the field just above our shelter. This depression traps dew and heat, the wind passes over the top and even in the winter there are soft wide blades of grass. It's surprising she ever leaves. She probably wouldn't if it weren't for the foxes and the need to mate, though if the Chinese were right and a female hare is impregnated by the touch of the moon on her back, she might not even need to leave for that. When we first saw her, she was stretching in the moonlight with her back legs lasciviously apart.

But she does leave – nibbling among the sheep on the higher fields and dancing along wall tops. She goes always clockwise. If I get in her way, she'll run straight up or straight down before resuming her rotation. Nothing, including a risk of death, will make her go anticlockwise. She never goes even into the shadow of the wood. I follow her for several nights, watch her ears for several days and learn to love her before I decide to kill her.

I don't want to kill her, and because I don't, I think I'm morally justified in doing so. I've killed a hare before. I lay in the flooded furrow of a turnip field in the Somerset Levels and shot her in the face as she lolloped towards me. I didn't just feel guilty; I felt unsafe. I looked over my shoulder for months. Now I'm planning to kill again.

This time I *want* to feel bad. We're here too casually. We know we can leave at any time. The grumble of the road makes a nonsense of our experiment. But one sure-fire way of making us feel part of the wood, part of its real story, and accountable to it, is to kill. The wood won't *let* us kill unless we are part of it. And to kill will require an act of contrition. This isn't some sort of perverted Dostoevskian thirst to know how killing feels. I know exactly how it feels, and I hate it. The hare isn't going to die for this book, but because Tom and I and the hare are all part of this place and because I'm hungry and because the hare is going to be eaten anyway, sooner or later, just like me.

We plan it carefully. The hare will be out and about until

dawn. She'll come in on the western side of the hollow: there's a clearly marked track through the grass. A spare hawthorn sprouts from the other side. Two humans can sit comfortably in it for hours, and the hare will have to come under us to lie up in her favourite place. She always comes slowly and cautiously, thinking of the fox. Our scent won't leach down the tree to her, and it will be easy enough to drop a stone right on top of her, following up with a throwing stick if that's not enough.

We've rehearsed it all. We've got a stone each, and we're going to sit on different branches. That covers all possible approaches. Tom has practised for hours with his stick, and he's deadly. He infallibly hits a small target at twenty yards, and from the tree he won't have to throw more than five. We can't miss, and we're planning what to do with the body.

Tom's disgusted by our lack of authenticity, and he thinks the hare can help put us right. From the start he said that we shouldn't wear modern clothes, and wanted to stitch skins together. I resisted, saying that that would be play-acting: that this isn't about physical emulation of Upper Palaeolithic people, but about entering their mental and spiritual worlds.

'But you're always saying that what happens to your body affects everything else, Dad,' rejoined Tom, astonishingly. 'I've heard you boring on about mind–body–spirit unities, and if that doesn't mean that we should wear skin trousers, I don't know what it can mean.' And I thought that at the dinner table he just daydreams about his bike.

Of course I had no answer, and so of course I gave a long, complex and pompous one, and off we went to Derbyshire swathed in fleece and nylon. But the hare's imminent death renews the debate, and we agree that its skin will be a foraging bag or the back of a jacket for Tom, its femurs flutes, its dried intestines the start of a cagoule (inspired by Inuit seal-gut garments), its ears (which we'll fray at the end) brushes, its eyes marbles, its shoulder blades knives, its ribs needles and toothpicks, its bladder a doll's purse, its feet talismans, its skull on a pole outside our shelter for some reason that neither of us can well explain, its vertebrae strung on nettle-fibre string as a necklace, its tendons the string of a fire-bow, its brain a paste for curing its own hide, its buttocks and shoulders roasted, its livers, kidneys and pancreas fried, its lungs bait for crayfish, its body boiled for soup, its dung used to fertilise a plot which will give us flowers in the spring and edible seeds in the summer.

Only if nothing is wasted can this death be excused. It is terribly certain that any waste will bring retribution. We don't know if that means broken legs or broken sleep or hauntings or diarrhoea, but we know it will be grave, immediate and lasting.

It rains on the day appointed for the death. We shiver miserably in the tree, hanging on like two old damp crows. I'm glad. We're about to do something dreadful, and it is sacrilegious to kill comfortably.

The hare plays her part in the cosmic drama beautifully. She approaches the tree right on cue, picturesquely, shining under a crescent moon. She pauses at the lip of the hollow, raises her head, scans the air with her nose and comes slowly towards us.

She is right under me. I can't miss. I drop the stone. I miss.

The hare, confused by the thud next to her, comes on. She is right under Tom. He can't miss. He drops the stone. He misses.

She bolts back up to the lip and pauses, trying to work out what's going on. No fox, buzzard or owl drops things from the sky. She's a few feet from us. There are no branches in the way of our sticks. We raise our arms and let fly. We can't miss. We miss. She makes off into the field.

X laughs.

I'm intensely relieved. We're not ready. Liturgically, I mean. I can feel X's stern propriety – a propriety he learned from the wild. Deaths are subject to a priestly code. We'd vaguely felt that the wood and the hare were under its jurisdiction, but its choreography is clearly complex and vital. To miss or fudge a step would be dangerous.

★

We sleep well that night, for the first time, despite or because of our hunger: because the wood is now our home. Our failure means that we're part of it; watched by it and subject to its rules. If we'd been modern colonials (like me), dropped off with our well-oiled shotguns at the end of the lane, we'd have killed that hare, eaten her rump merrily by the fire (with a nice claret, no doubt, brought up in a hamper from the 4 × 4 by a wife-slave), slung the rest into the hedge and thought that we were fine manly things, in tune with nature and at its summit.

Would we have escaped the wood's retribution? We would not. That hare's outraged soul would have clamoured for justice and rest, would have jumped onto the leather passenger seat and paced the fitted kitchen and (especially) the deep pile carpets of the bedroom until it had achieved whatever was necessary. We probably wouldn't even have realised that we'd been judged, sentenced and the sentence executed. That ignorance would have doubled the severity of the sentence. Ignorance of the law of the wood is a serious aggravating feature.

★

We try to learn the liturgy: the way to do things properly; the way to avoid offending the fastidious, prescriptive and vengeful guardians of the place. Everything matters. We watch the rain fall on one leaf, trace the course of the water under a stone, and then we go back to the leaf and watch the next drop. We try to know snail slime with the nose of a bank vole, stamens with the visual resolution of a bumblebee and the leaf pennants on tree masts with the cold eyes of kites. We see the patterns of lichen on the rocks and shout at ourselves: 'That is not like a map of New Zealand. Nor is it like a plane. It is not even like a bearded man, or a wolf. It is just what it is.' We outlaw similes when we speak, because if we see things properly, nothing is at all like anything else. With that self-restraint, metaphors paradoxically ignite. Things aren't *like* other things. But they can *be* other things. A tree doesn't roar like a lion in the wind, but it can *be* a lion.

We keep the wood's hours and fall in with its moods. When it rages, we shake our fists at the sky. When it is sullen, we sit on stumps and look petulantly into the middle distance. When it is sad, we stroke it and ask it to stroke us. Anthropomorphism? Definitely and precisely not (though anthropomorphism is a very underrated experimental method). We're not making the wood in our own image: projecting ourselves onto it. It's making us. If we let it, it would do a thorough job, chafing off our pretensions, waking us up, melting the divisions between the parts of ourselves and between us and other things; harrowing and planting us; making us green and interesting; giving us new and better names. But we won't let it. I don't have the stomach for such costly healing, and Tom has to go back to school some time, says the State.

There's a runnel of dirty water crawling out of the ground, trying to be a stream. We humour its ambition by sitting with our feet in it, and it rises to the occasion and sings.

We know that the trees are connected by a vast, dense and sensitive plexus of fungal mycelia. It makes the wood a single organism. To tread on one part is to tread on all parts. To stamp on one part is to abuse all parts.

We try to walk lightly. We can't feel the mycelia fizzing under our feet as the reports of our behaviour dash from tree to tree, but we can sense in the leaves and the bark what the reports say, and are shamed into decorum.

We try to slow down to the pace of the trees, and speed up to the pace of a wren's heart, for there are many parallel time zones in this wood, and if we can navigate through them all we might be able to taste time itself, or – which comes to the same thing – escape time altogether. We try to slow down the birdsong in our heads, so as to hear the lugubrious melody encrypted in a squirt of noise.

Tom thinks he can hear small mammals breathing under our heads as we lie in the shelter. I don't doubt it. But when I praise him for his sensitivity, he denies it. 'Don't be daft. No one can hear that.'

'But you said ...'

'I was joking.'

No he wasn't.

We watch clouds, fires, the metallic joints of insect legs, the layout of bird entrails, the single leaves that flap on tree tops when the rest of the tree is still, the bite a caterpillar took at the edge of a leaf last summer just before it was eaten itself, a seed that had forgotten to fall and another seed ground into the earth by Tom's bare foot.

We remind ourselves that we have unused noses, that scent is always surging in unnoticed and that our scent is always pouring out. We pay attention to the smell of the cold itself, which is a primary scent, just as blue is a primary colour that cannot be resolved into any components, and to the courtier scents that accompany the grand sweep of winter into our noses: the citrus slime of snails, our own shit, lanolin, mould in worm tunnels, the different, dustier mould of sour and mourning air and the seaweed edge of the wind when the pale sun gets to it.

But instead of the wood coming to us, my father comes. This doesn't mean that we're failing to learn the liturgy. It means we're learning fast.

My father felt a connection to this place. We lived not far away, and he paced the woods whenever he could, and wrote me very formal letters in his old-fashioned copperplate script describing where he'd been. When I had exams, he sent me bags of talismanic leaves and twigs and cones. I put them on my desk as I sat the exams. We took it for granted (though we never found the words to say it to one another) that there was some wisdom in the land which flowed through the leaves and into my pen. I have one of those bags still. It's on my desk now.

We burned him, as we burned my mother, in a municipal furnace hundreds of miles away, powered by the bodies of ancient sea animals from somewhere under Arabia. We hired strangers to put on black suits and solemn expressions and to drive him from a breeze-block garage in a shiny car with a smooth automatic gearbox, though he was only ever happy travelling in his battered Land Rover, and we hired a vicar who never knew him to say kind things based on our notes and a remote theology.

Nothing – not even my laptop, my reductionism or my central heating bill – exposes as pitilessly as that how far I'd slipped from the world of the wood – X's world. I can see X now, shaking his head in disgust as he sees what we did to my dad. We should have stripped off his old tweed suit and carried him in procession to a platform at the top of the moor, laid him down and let the crows take him. That's what civilised people did hereabouts. Or (as they'd have done a few thousand years later, when settlement happened) buried him right under the hearth, because to leave the hearth was, unthinkably, to leave the family. Then the children would have grown and played on top of him.

He was happy up here. It was part of him. Now that he's burned, though the burning was in Somerset, some of him is literally part of this place. He was metaphorically inspired by the trees. Now they have actually inspired him. Bits of him have gone through their stomata and been built into their cell walls.

But he often forgot he was part of the place. He watched TV for months on end before remembering that it made him wretched, and then he headed out again to the hill with a plastic collecting bag and wrote one of those letters, and smiled again as he smiled when I was very young.

Though he loved it here, and loved me and Tom, and though he's gone through the pearly gates of the stomata, it's a surprise to find him here. Perhaps he's protesting against the barbarity of the automatic gearbox, or trying to point out lucky leaves for Tom's exams, or perhaps he just wants to go camping with us. Or perhaps he's applied successfully for the job of liturgy teacher. That, after all, is a typical job for a dead Upper Palaeolithic man, and perhaps we all become Upper Palaeolithic after death, when all the poses of the ages melt away with our melting flesh. Why ever my dad is here, his accessibility encourages me. It suggests that the walls between things are thinning. If we're to kill safely, we'll have to dissolve walls: to go to the hare and ask permission to crush or spear her, or (more advanced stuff, this) turn ourselves into the hare, so that a rock on her head is a rock on ours, and therefore not morally reprehensible.

He was a teacher, my dad. He never stopped being one, and I doubt he's stopped now. He had a chalk-stained jacket with leather elbow patches, smelt of coal tar soap, wore a National Trust tie, always wrote with a fountain pen and announced in school assemblies that he'd be grateful for animal corpses for his son to stuff. He always had a load in the back of the car when he came home, and he'd sit in my taxidermy shed in the evening, telling me about each organ. He was great with wood – he was an inspired carpenter and carver – but he couldn't be doing with metaphysics. Most of him was Irish, though, and he cried easily and didn't doubt that there were Little People under the hill. He'd have seen X long before we did, and would have shared his old pipe with him and known his life and death history by now.

My dad's occasional stiffness, I see now, was choreographic. Like X, he knew that the world, being the way that it was, demanded a particular approach: that reality would grant no audience unless you dressed properly, washed behind your ears and knew how to walk.

Those animal bodies in the car were sacramentally important to him. No doubt he foresaw his own dead body in them. He was often nearly killed running into the road to pick up a dead badger to stop it being mashed. 'It's not dignified to leave him there,' he'd say. 'I've put him back in the flowers, where he belongs.' That's pure hunter–gatherer. I can see that he might teach, *post mortem*, what he now must know even more authoritatively – about the nature of things.

★

'Dad,' says Tom. 'That place by the alder trees, where those two jays were fighting. Where you sit in the summer to read: it always smells of coal tar soap.'

'That's not surprising,' I reply. 'It's pretty marshy there. What happens is that all the dead vegetation rots anaerobically and produces marsh gas: hydrogen sulphide and methane and other things. That's what you're smelling.'

★

There's only so much you can do by staring, sniffing and thinking about your dad. Remember the arduous journeys of the pierced shamans – journeys from which the shamans returned bearing knowledge of herd movements and, much more repercussively, perspective: knowledge of themselves: *consciousness*. Those journeys were painful and terrifying. Consciousness may have swept into human history on a tsunami of vomit from plant entheogens and that, as any modern narconaut hanging over a bowl in a Peruvian hut during an ayahuasca séance will tell you, isn't nice. Enlightenment didn't come cheap, by reading a self-help book or by talking about God over pizza. You were torn apart, or crawled into the earth's belly through a passage barely bigger than the birth canal.

Years ago, after days of fasting in a peat hole on a moor near here, I sprouted black wings, flew over myself and the moor as I had hovered more suburbanly over myself in the hospital, cruised above a road by the reservoir, croaked, ate a frog and stuck my beak into the chest of a long-dead sheep.

It scared me rigid. I didn't want to go back there, and swore I never would. I later learned enough about real shamanism to think it arrogant to call my crow experience shamanic. The real thing is *much* scarier. I have no truck with the easy, cod shamanism that makes shamans sound like teddy bears. They're real bears, with red teeth.

Every time I read something new about shamanism I renewed the vow. I wasn't going anywhere near it. Yet here I am in the wood, convinced that shamanic experiences are central to any inquiry into human origins. I remember with a jolt that the foundation myth of Christendom is the tale of a pierced shaman who shuttled to and from other worlds, bearing, it is said, great gifts, and doing great acts on behalf of the people. In a few days' time there will be a feast to commemorate the first of those journeys, as the polite bells of the village church sometimes recall. So I won't escape from shamanism when we get the train home. Far from it: I'll be catapulted into a celebration of one of its most explicit instances. And now the wood is constantly demanding more of me: clutching at my face in the day and leaching the heat out of my body at night, whispering, either as a threat or an expression of solidarity: 'You're part of us. Come further in.'

I'm no hero. I park the thought of shamanic journeying. Or I think I do. I assume I'm driving.

Perhaps X is a shaman or an incipient shaman. Perhaps that's why he's here with his son. Shamans are always on the edges. They live at the edge of settlements, linking the people to the wild, and at the edge of ordinary consciousness, negotiating deals with the spirits on the other side. If so, it's natural enough that X should be so confidently here on the other side of his own grave. Perhaps he's trying to teach the boy how to fly, or the language of dandelions, or how to persuade caribou to run over a cliff, or the right way to greet an animal you've killed, or the position to adopt when you meet Mars, or how many psilocybin mushrooms to take if you need to hear the colour of a wolf's throat, or how to pull a wounded soul out through a nostril, bind it up and push it back. But mostly the lessons will be geographical: the routes of paths through the stars, the roads taken by the dead and the fissures through which evil wells up from the ground.

I think the son is overawed. He'd rather throw stones into the pond with Tom, and juggle with sticks. But if his father hasn't caught consciousness just yet, the boy should beware. It's highly infectious.

*

We wake with the lean sun, do the rounds with our mammals and birds, fill our hats with the last of the berries, set optimistic nettle-string snares, knap hand-axes from the flint we've brought with us, raid the roads for meat, sleep and pick at our feet.

Time passes in its own mysterious and erratic way. I don't have a watch even at home, and have hidden the clock on my computer. Why would anyone *choose* to be subject to the totalitarianism of the clock when it's so easy to be free? Tom has left his watch behind too, and watches the strange behaviour of time with the same fascinated attention he brings to woodpeckers, rabbit claws and fly eyes.

'Today seems to have started yesterday.'

'That cloud stood still for a day and then decided to catch up.'

'Everything in this wood is old, but nothing ever gets older. I wonder if we'd live for ever if we stayed here?'

'Dad, what decides when the day ends? The sun or the stars? Which is most important?'

At night we breathe fog and smoke, and cough with the sheep. The wet wood screams in the fire. When we lie down, all we can hear is the slap of the tarp and the hiss of the fire and the roar of the trees.

I'm beginning to think that no dramatic out-of-body experience is needed for the kindling of consciousness. Lots of staring into a fire will do instead. A fire turns literal creatures into symbolisers: it makes everyone a metaphorical and story-telling animal. A fire creates, and shows how the creator also destroys. It confounds the boundaries of matter. It makes gas from fluids and from solids. It eats wood, sleeps and is woken by human breath. It makes a nonsense of space. Though it can travel (as Tom and X carry it) as a tiny dark spark inside a black fungus ball, it can fill a forest. It births metaphors. And there's no political philosophy that can't be deduced from staring into a fire. Logs won't catch without small twigs. Small twigs are both the death and the apotheosis of logs.

Metaphor, too, could be a creature of fear. Our software is programmed to see snakes when they're not there, for very obvious reasons: better a string of false positives than one false negative. One false negative might be your last. Imagine, then, that you're walking through the bush back to that night's camp through the half-light. You're about to put your foot down on a rock, when you see that there's a puff adder there. You just manage to avoid it. Breathing fast and gratefully, you look back at the snake. It's no snake at all – just a fat tree root. How can you help noting to yourself that a tree root can be like a snake? There's the simile. Simile and metaphor are not the same, but in the case of the snake–root they're not so far apart. Sitting round the fire that night, mightn't you easily think that a root can stand for a snake? The symbolic revolution would have kicked off in your head. Once it had started, it would be impossible to stop. The nerdish, literal left brain would protest, of course. 'It was just a root,' it would whine. 'Snakes bite; roots don't. They're completely different.' But this sort of peevish pleading couldn't prevail – or at least, it didn't until the left brain's coup was nearly complete. The bush was suddenly full of story. The bush burned, as it did much later for Moses. Things were suddenly not just as they seemed. Every object had a hinterland; everything *signified* – at least potentially. The world was infinitely bigger, more colourful, more complex and more relational than it had been.

Like everything, this started in Africa. Red ochre was being used there from at least 100,000 years ago, and in South Africa's Blombos Cave there's a piece of red ochre that was incised with a very deliberate pattern 70,000 years ago. Perhaps we shouldn't get too excited about the mere presence among human traces of red ochre: it was useful as well as symbolic (it has been mined for at least 300,000 years, and is used in glue, as a lubricant, and for preparing skins), but it is safe enough to see the incised ochre as at least a dim flicker in the fire of symbolism that spread north and eventually everywhere, setting brains ablaze.

There aren't many other obvious early traces of the fire in Africa. There are some pierced shells in Morocco from around 70,000 years ago, and some grave goods in early Palaeolithic burials in Qafzeh and Es-Skhul in Israel from around 130,000 years ago. No doubt there is much more to find in Africa, and no doubt it's wrong to say that the first behaviourally modern humans were Europeans, but even accounting for the Eurocentric bias, something special happened in Europe from around 50,000 years ago. The impression is of little flames creeping up the Nile Valley, turning east through the Levant and up through Anatolia, and then hitting a mass of tinder-dry neurological brushwood in Europe. It is fanciful to think (as some have done) of a symbol-facilitating genetic mutation arising in Europe at this time. It is better to suppose that there was an incremental accumulation of symbolic tendencies, and that when that accumulation hit a critical demographic point – whoosh!

*

We don't try to kill again, for the moment. Instead we scavenge, which was an important means of survival for Upper Palaeolithic hunter–gatherers. Or Tom does. He's sleek on the fat and marrow of road-verge rabbits and pheasants. He carries roast legs in his pocket and crouches by the fire with spitted hearts.

I've stopped eating. I'm learning that a taste for clarity can be as addictive as a taste for alcoholic numbness or sex. I'm desperate for the quiet, shimmering clearness of the fast that comes after the first day, where everything drops hints that it might not be quite as it seems; that we see only surfaces; that if we turned our head a little we'd see that that grey birch bole makes a peacock's tail look like an ascetic's cassock.

★

After three days of fasting the pain has gone. I'm tired and comfortable. I don't notice the cold much. Tom's not fasting, and goes off alone on his foraging trips. I lie still for most of the day, watching; waiting for something to happen. The shimmering comes.

Waiting, without a screen, a game, a person or a sensation to entertain, might sound like hell. But it's so exhilarating that I can't stop panting, and can hardly keep quiet. I shake with the shimmer, and know that something is coming that will change everything.

There's nothing extraordinary or difficult about fasting. It's easy. You just stop putting food in your mouth. For most people, for most of human history, it was as much a part of life as defecation, and about as necessary for health. Being able to deal with an empty belly was far more useful then than being able to type or drive is now. Like wolves, we're built to glut and then starve. Regular meals are deadly. Cells that go hungry live longer. They can sometimes even *become* younger. Perhaps that's what's behind the shimmering. Perhaps a younger and younger me is watching the sky. You'd expect that to produce interesting sensory effects.

★

Because I've not been eating with him, I haven't been noticing what Tom does for food (other, dear social workers, than ensuring he's got enough), but I begin to see that before he eats at the fire, he goes off into the wood for a few minutes. I say nothing. I'm obviously not invited. But it makes me wonder, and one day, when he's away up the dale, tracking roe deer and trying to perfect his jackdaw calls, I walk in the direction that he goes before eating. There's a well-trodden path.

It takes me through a birch glade, over a wall, and through a thorn arcade into a clearing. There's a big stone there, dug out and dumped by the miners. On the stone are small bits of bone and mouldering meat.

★

Snow comes out of a taut and sickly sky, lightly at first, and then not so lightly. The tarp sinks under its weight. The snow brings geese, and a new kind of crystalline silence. There are different kinds of silence, like different kinds of snow.

If we had proper eyes, we'd have seen tracks before. There's no way of missing them now. The snow freezes movements: times are compressed together in one place. Tuesday and Friday lie side by side, to be read together and interpreted in the light of one another. This is how time is supposed to appear to God, sitting in an eternal present.

The snow changes the routines of our guide animals. The rabbit sits at the entrance to his burrow, sneezing when snowflakes land on his nose. The magpie doesn't go to the fields, and doesn't tick-tock so much. The robin has shaved the base off his pentagram. And the hare won't leave her hollow. Her eyes are barely above the snow line, and the snow covers her back. She must see nothing but white. Unmoving, undifferentiated white for six days and nights.

★

'If you have to play charades,' said the highly scented woman, 'you really should leave it to the summer. Or at least the spring.'

She was wrong about that. This is far and away the most important time to be here. Winter has to be the first and formative and longest chapter of this book, for it was the first and formative and longest chapter of *me*. Early Europeans were creatures of the dark and the ice. We can only avoid madness in the teetering glass boxes in which most of us live, work and breed, because they remind us of the towering glaciers that were, after the thorny bush of Africa, our second home. We were conceived under an African acacia, and Europeans then grew up in an ice cave, with a woolly rhino grunting outside.

We look always to summer, thinking of winter as simply to be endured, but winter is when our sustaining fables sprout; it is when humans huddle together, making the relationships (and hence the differences) between them more obvious. Relationship and individuation both flourish in the dark. And the dark is more *other* – full of teeth and hair.

It's commonplace these days in certain circles to say that we're part of the natural world. That's wholly true and wholly untrue. Certainly when spring comes you can think that you're simply part of the wood. But no one thinks they're part of a wood that snarls. It was from the tension between the whole truth that we're part of the natural world and the whole truth that we are not that human consciousness erupted, flowed out into the ice fields and solidified into the stuff from which we're now made.

I suppose there might be some comfort in these thoughts when we're reflecting on the fact that one day we will go back into the dark and cold. Or perhaps not.

★

I'm worried about Tom. He's becoming a disciple of the sinister Bear Grylls. He's more interested in the technology of survival than the point of survival, and he talks in Bear Grylls's imperial language: about conquest and triumph; about the wild as an adversary that has to be tricked or crushed. This is anachronistic language – the language of the Neolithic (we'll hear its accents shortly), refined and made diabolical by Western Christianity's mis-exegesis of the Genesis mandate to subdue the earth. Mountains can't be conquered. The best you can hope for is that the mountain, on the day you choose to climb it, chooses not to kill you. If you're shrewd, you'll spend time and effort trying to persuade it not to. We've long since given up our efforts at propitiation, and (who can now deny it?) the earth is rising up angrily and is about to fall on us.

Tom's technology (flint knives, scrapers and axes, plant-fibre rope, pitfall traps, spears with tips hardened in the fire, spear throwers that double his range, his force and his accuracy) are becoming not instruments for the performance of a sacrament but weapons in a war with the wild: part of a wall that separates him from the wood. This, I suppose, is a recapitulation of what actually happened in human history: a warning about how the things we use come to use and to change us.

But for most of the time Tom's my door into this place. He's still a child. The process of knowing anything about the world is the process of *anamnesis* – unforgetting – and he, like all children until we corrupt them, has forgotten much less than I have. Then there is his sensory superiority, and hence his freedom from cognitive tyranny and the malignant distillation of glorious trees into dowdy propositions.

The bonds of language and proposition that prevent my mind from stretching out and touching the real world are loosening. It helps not to speak. It helps to breathe consciously. But it's slow work. However linguistic the Upper Palaeolithic inhabitants of this wood were, I'm sure that they used language as a tool rather than (as in my case) allowing it to build and to constitute an entire virtual world in which they were (as I am) forced to live.

★

I saw X again last night. He was leaning against a tree as the last light left, and looking straight at me. I couldn't see any expression on his face, but after a while he lifted his hand in some sort of farewell, turned and walked away. And then I saw that he walked with a terrible limp. His left leg is twisted.

★

Tom returns with a squirrel. He revives the fire, roasts the squirrel and then sits looking into the dark rather than the fire, which isn't like him. When the eye of the big bear has climbed above Sarah's house and is squinting down at us, he starts to talk. He doesn't want me to answer. He's troubled by the idea that healers have to be wounded themselves. I can't help him. I'm troubled too.

He turns back to the dark.

★

I've not eaten for eight days now.

'You're not quite as porky as you were,' says Tom. He's right. My cheekbones are cutting into my face, and the border between me and the wood is thinning. We're oozing into each other. My shape is shifting all the time.

The shape I call my own is created by the pressure of the other entities, both human and non-human, all around me. Take away my relationships and I stop existing. You can't describe me except in terms of the nexus of relationships in which I exist. You can't say where I am except by triangulating from the positions of the other creatures in my world.

X knows where the other creatures are. He therefore knows where I am better than I do. He knows there's a roe deer lying in the thorn thicket, a jay at the top of the oak and a wolf running down the Pleiades just over my head. He sees that very little of me is here, attending. He sees how Tom shifts uneasily from foot to foot when I talk to him; I no longer believe that language corresponds at all well to the things it tries to describe. And so it can't stick things together: make parts into a whole. X lived in a whole place. His bits of tundra joined up. What can make our wood whole?

The obvious candidate, as I lie looking at the snow dancing down to an invisible tune, hearing the symphonic roll of the wind, watching the quadrille of rooks in the field, and the circling of the stars, is music.

Later I wrote in panic to the biologist David Haskell, an expert on birdsong, begging him to reassure me that music is 'chronologically and neurologically prior to language'. It surely is, he replied. 'It seems that preceding both is bodily motion: the sound-controlling centres of the brain are derived from the same parts of the embryo as the limb motor system, so all vocal expression grows from roots that might be called dance or, less loftily, shuffling about. An epistemology that is grounded in muscle, nerve and bone is perhaps, then, what we need.'

Yes! That epistemology is urgent. Not just to understand the Upper Palaeolithic but to understand *now*. And to live well as humans, for how can we live well in a world about which we know nothing or (because language has fed us fake facts) less than nothing? So I'm going to dance and walk and run all the lies away: I'll exorcise them with my tin whistle and the B Minor mass and rebetiko from a smoky cellar in Piraeus.

I'm also going to use more than one of my five senses (surely there are many more than five). I use only my eyes. If there are only five senses, it means that at best (and not allowing for the distortion that comes from the unholy trinity of vision, cognition and language) I'm getting only one fifth of the available data about the world. Just imagine making decisions about business or relationships based on only 20 per cent of the relevant information. We'd be bankrupt, and our relationships would be disastrous. Yet that's what we're doing with the *whole world*! Our intuition is older, wiser and more reliable than our underused, atrophied senses. We intuit that the world is one way: our senses insist that it is another. There's a sickening gap between the two types of understanding. No wonder we don't feel at home in the world. We have no idea what it's like, and we *know* at some level that we don't know what it's like. Just imagine how much more satisfactorily and intensely I'd inhabit this wood if I got just another 20 per cent – by paying attention to the scents surging constantly into my nose, for instance. Just think how much more I'd squeeze out of life: how much more living I'd do. X, like a fox, used all his senses. If he didn't, he'd have died long before he got that beard.

That's work for the future. For the moment I have to make something of the shimmering: to observe, without resenting, the protests of my stomach and, most of all, to be some kind of father to my son.

★

Tom is clear: if we're to do this properly, we need a dog. He's looked it all up. Wolves may have been domesticated 40,000 years ago, he says, and a tame or tame-ish wolf in this wood would change it massively. He's right about both of those things – though I doubt that dogs reached this part of the north European ice field anything like that early. He has in mind something like a big lurcher that you could mistake for a wolf in the half-light. We could easily borrow one from an Exmoor friend.

I resist.

There are no domesticated wolves in England; there are only mere dogs. And these dogs have co-evolved with us: for us. We have shaped them, sadly, far more than they have shaped us. Studying dogs is more anthropology than zoology. (Perhaps that has very recently changed, and they're now beginning to drive our evolution rather than us driving theirs. I hope so.) They bear in their bodies and their psyches the post-Palaeolithic history of man. Dog and human co-evolution accelerated fast from the Neolithic onward.

If you want to know how far we've moved from the place we were designed to inhabit, look at modern dogs. The tragic, wheezing ones with bows in their forelocks, and squashed faces and bent legs. Proper dogs – the ones with faces like wolves – look at you out of the distant past. But the past from which they look is not distant enough to make them right for the wood.

A proper dog would, of course, be closer to the Palaeolithic wood than we are. That would be another problem. It would take over. The wood would react to it rather than to us, and we'd react to the dog rather than to the wood. We're supposed to be studying humans, but if a dog were there we'd study the dog instead – and a modern dog at that. The dog would insulate us from the wood. We'd perceive it through the dog's senses rather than our own. I tend to travel abroad alone to avoid this vicariousness.

So the lurcher has to stay on Exmoor.

★

The snow is still falling. At home they will be worrying about exposure, frostbite and surrender. As the contours of the land are smoothed out, the land becomes spikier and more dangerous. The birds have a new desperation. It's harder to start to flutter, but once they've started it's harder to stop. In the summer their flight across the wood is easy, fluent and mellifluous. Now it's frantic and staccato. In the quiet I translate everything into music, and there is only dissonance. The rook clashes with the robin, the fox with the wood pigeon, the hare with the tree.

In the cold is the knowledge of personal extinction or transformation: of the journey across a border to wherever my father is or isn't. I hug Tom tighter.

The robin and the magpie come even closer. It pleases me to think that they want our company more than our food. We are part of their old, kinder world of brown and grey. The rabbit scratches through the snow to get at the grass, as X's caribou did to get at the tundra lichen, and one day leaves his ears and a foot by a wall. Only the hare and the fox are poised in the cold. If I lie at the rim of the hollow, I can see a tiny trickle of steam coming from the snow. It's from the hare's nose. She lies in a snow cave moulded by the heat of her own engine.

We too try to turn the snow against itself, and build low banks of snow at the sides of the shelter. They keep off the wind, and make us feel that we're doing something about the cold.

There's plenty of food for us. Pheasants are drunk on the cold, and don't bother to get out of the way of the cars. We're warm enough too. Before the snow came we manically gathered wood and stored it under the tarp, and its community of woodlice and spiders has been joined by other refugees. There's a long-tailed field mouse. We can see the end of his tail twitching as he sleeps or flinches between the sticks. The tail makes the shelter home for us.

Our fireplace is the heart. It keeps heat pumping through us. There are always at least embers there. When Tom comes back from a foraging trip, he says he'll wake the fire. I haven't given him that thought or those words. In fact, he seems to think that a fire doesn't even need embers to exist. It's always there in a log, waiting to be woken. It's as if the fire is the soul of the firewood – a special case of the activating, defining sparks at the core of everything.

He won't let me tend the fire. It's his business. His movements around it are the slow, careful movements of a priest around an altar. When I suggest moving our shelter to a drier place, he won't have it, 'because we can't really move the fire'.

His instinct is right, but in fact the ancients, fully accepting Tom's theology of the home-fire, found ways of moving house. When Aeneas carried his aged father, Anchises, out of the flames of Troy, Aeneas' son, Ascanius, was at his side, carrying the fire. The fire was sacred. It *was* the home, which itself was sacred. Wherever the fire was, there the home would be.

Virgil was recording something with origins very much earlier than the Homeric Bronze Age, and very much more fundamental to human identity than a Romanised trait with an eastern Mediterranean ancestry. X had a home in his fire-bag. Home smouldered in a ball of moss, waiting to flare. To wander was not to be dispossessed.

Anchises had a pot in his hand. It contained the ashes of his fathers. They defined home just as much as the fire did. Ancient humans always had their dead with them – in a bag, under the kitchen floor, around a wrist, on the mantelpiece or at the head of the table. We stuck our parents' ashes in shiny boxes (made in China) that we chose from a glossy catalogue. We splashed out for my dad, and he got brass corners. Then we dumped them in a wet hole in Somerset.

And we wonder why we feel we've lost our grip on fundamentals. After the ravens had picked my dad clean, I should have put his skull in my backpack, made a necklace of his teeth and used his pelvis as a pillow.

★

The winter is the time of palpable edges: viable/doomed; black/white. Tree spikes stab the wind. In fact, everything in the natural world is about edges. If your face is near enough to the ground anywhere except a sprayed, antiseptic city park or the field of an industrial farm, you'll know that to move an inch is to move between zones as different from one another as Antarctica and the Amazon. If you know this, a walk through the wood is travel fantastically more exciting than anything you can do in a private jet. Each step is a journey across many domains and frontiers. Yes, there is a whole, but a whole that's complete only because of the vibrant individuation; because of the edges. The real world isn't mushily homogenous. It has no monocultures. X never walked on the same thing twice. *He'd* never say 'We're walking on grass', but 'This toe is on *those* blades, and this toe on *those*': he'd name the fifteen grass species under each foot, beg forgiveness (possibly in their own languages) for crushing them and thank them for cushioning him.

All change – *all* change – has always and can only ever come from the edges. Nothing of any significance has ever come from the centre – from parliaments, from Cabinets, from boardrooms, from think-tanks with the ear of ministers. Evolution needs edges. Turn the world into a monoculture and you'll have shorter lengths of edge, and thus less change, and thus less evolution. That's bad news.

★

Suppose that behavioural modernity began 40,000 years ago, that the Neolithic began 10,000 years ago and that we became modern, in the sense we now are, 1,000 years ago. (As we'll see, I argue later in this book that this last transition was rather more recent than that.) Assume that each generation is twenty-five years. There have then been 1,600 behaviourally modern generations, 1,200 (75 per cent) of them Upper Palaeolithic or Mesolithic. There have been forty modern generations: that's 2½ per cent of the total human generations. If a human life time is seventy years, 75 per cent of a human life time is about fifty-three. Most of our development as individuals is done by the age of fifty-three. And most of our development as humans was done by the end of the Upper Palaeolithic. We're Pleistocene people.

It's laughable to consider modern humans as normative. We are recent and palpably inadequate mutations. But take heart: we can reverse the mutation.

If we start the story when anatomically modern *Homo sapiens* first appeared, 200,000 years ago, 95 per cent of our history has been as hunter–gatherers: as edge-creatures and hence, excitingly, as changing and change-bringing creatures. Now most of us are in the centre (of cities, of movements, of presumptions) and so have stopped changing either ourselves or the world in the way that the early human generations did. We think that we're in a fast-changing world. Well, perhaps, but humans aren't changing in the way that the Upper Palaeolithic changed us. What we think of now as change is angst and dissolution. The changes on our watch are not the multiplication and refinement of nuance or the deepening of understanding. They are acts of vandalism: the spoliation of things, places and modes of being that are ontologically superior to us.

There: I feel a lot better for having that off my chest. Same time next week?

★

The expectancy of the early fast-days has been replaced with vertigo. I'm on a ledge, it's a long way down, and dark, and I don't know what's down there. We're at the edge of respectability, and get strange looks when we stuff roadkill into our bags and traipse back up the dale past the pub.

The last thing I ate was a hedgehog. That was nine days ago. From the taste of them, hedgehogs must start decomposing even when they're alive and in their prime. This one's still down there somewhere, and my burps smell like a maggot farm. I regret its death under the wheels of a cattle truck far more than its parents or children can possibly do.

We're being changed by the edges of this winter wood. We're at the edge of the village and spend most of the day and night at the edge of the firelight. I'm starving, and at the edge of my experience of fasting.

We're at the edge of our understanding of the kind of creatures we are, and near the end both of our respective tethers and, thank God, the time we can spend here. Soon I can get back to calling myself a nature writer by sitting at a desk, looking soulfully at clouds and listening to electronic birdsong.

The most significant edge for me at the moment is the edge of sleep and waking. I spend a lot of time in no man's land.

Why didn't I think of it before? If you want to know what it was like at the dawn of consciousness, well, watch consciousness as it dawns. Sometimes literally, in the very early morning, while light and dark weep into one another. Watch your own waking. Wakeful dreaming is an important spiritual discipline in many religions, and now I know why. It's a powerful lens for the examination of your own consciousness.

When light and dark meet, and when consciousness and subconsciousness meet, there is no misty twilight. Nothing is amorphous. On the contrary, this is a territory of unusual clarity. Its light is like Attica in October. It's unusual because reality declares itself to all levels of being: to palpation and apprehension, to finger and gut, to balls and brain, to the thin convoluted rind of the cortex and to the deep, thick fish brain, built from phospholipids and bathed in sea water and full of memories of snapping plesiosaurs, to the part of me that knows the number of a wolf's chromosomes, and the part that likes to sit by the fire so that it can shove a burning brand in the wolf's face when it comes. Unusual because these levels are usually antagonists, but here they are friends.

I can hear their conversation, and know something of the relationship between them, simply by dozing and getting Tom to kick me when he sees me nodding off. It helps to be starving (which isn't nice), but even if you're starving, wandering in the no man's land between wakefulness and sleep is a lot less unpleasant and much less risky for your kidney function than the alternatives used often by our ancestors, such as eating hallucinogenic mushrooms.

The weary watchman at the campfire on the Derbyshire tundra, his head, under his elk-skin hood, dropping onto his chest, must have known this territory well. Indeed we all go there whenever we sleep out in the real world of owls and foxes and shrieking deaths, rather than in our soundproofed boxes. It's as if human sleep were designed to be broken. We are far richer when it is. I wonder if that's part of the appeal of camping. A muddy pitch in the rain between sleepless sociopaths with white vans full of white cider can only have hypnagogia to commend it.

★

Now listen (I'm talking to myself). The traditional way of looking at the birth of human consciousness is plainly wrong. There was consciousness long before the Upper Palaeolithic, and not only in humans. Many non-humans certainly have a sense of self: it's been conclusively demonstrated in primates, cetaceans and several species of birds (which of course don't have a neocortex at all – showing that consciousness isn't associated with brain evolution's most recent innovation, and that there's therefore no room for our habitual chronological snobbery), and many suppose that consciousness is ubiquitous in the natural world.

We're very bad at looking for consciousness. We tend to assume that if consciousness can't be indicated in the way that *we* indicate it – for instance by looking at ourselves in the mirror and pointing to marks on our face – there's no consciousness at all. But we're getting better at looking for it, and the better we become, the more consciousness we find. The universe seems to be a very fertile garden for growing it.

Yet it's clear that *something* tectonic happened to human consciousness in the Upper Palaeolithic – whether by revolution or revelation or evolution. A new type of consciousness emerged out of or in addition to or in substitution for the consciousness that had been there before.

Almost every human in the world, until the seventeenth century CE, assumed that the world as a whole, and every little thing in it, from pebbles to whales, has some sort of consciousness. I tend to agree. And religions east and west posit a consciousness possessed by the entire universe – perhaps, or perhaps not, equal to the sum of the individual consciousnesses present in the universe.

The relationship of this universal consciousness to the particular consciousness of individuals is a matter of abiding mystery and urgent importance. The account of the relationship that makes most sense to me is Iain McGilchrist's:

it turns on the relationship of consciousness to *matter*. Individual consciousness is what happens when universal consciousness is constrained in some way by matter. It's as if my body, like the pseudopodium of an amoeba, engulfs some universal consciousness. That bit of consciousness, for a while, takes the shape of my body. Bodies determine the way that consciousness behaves. We're not talking here about some crude Cartesian dualism, in which a soul that's the real me occupies my flesh like a squatter, telling everyone that my body is his.

There are many different types of body, and it's not surprising that (as we increasingly recognise) there are many different types of consciousness. A killer whale's body is not like mine (well, perhaps not *completely* like mine, as rude Tom would say). Its self is not like mine, and so its sense of self is not like mine. Not lesser, necessarily: just different.

In the humans of the Upper Palaeolithic a new *kind* of consciousness (or at least a kind of consciousness new to humans) arose, apparently without a significant change in the shape of the body. It's arising in X. That's the best explanation for the puzzlement behind his beard, and the panic on his son's beardless face.

*

The fasting, the shimmering, the edges of sleep, the edges of leaves, the edges of species, the edges of chewed bones and the edges of all categories are creating a new kind of consciousness in me.

In the moments just after Tom kicks me there are faces and patterns. The faces are usually kind but stern. At the edge of sleep I expected a giggling Flora with a basket of flowers and her tits out, but she never appears. They're not the faces of wraiths: they are, above anything else, *heavy* – from a world more solid than this one. Sometimes they're crowned or draped in leaves, and these leaves would cut through the stone of a Derbyshire wall as a flint knife through air. The faces never speak because they are before and beyond words, and so their eloquence is to ours as the leaves to the stone. They make me come over all Platonic or (if that's different) Jungian.

Behind the faces is a geometric matrix, or a sky full of regularly spaced dots, or a lush, vegetative world of tessellated ferns like they say you see on ayahuasca. The ferns sway, showing (as if it needed to be shown) how still the faces are.

Sometimes there are figures with peeling skin. Beneath the skin you can see what they really are. Sometimes they can be reached, or properly surveyed, only by flying. X is one of them. His brows are heavier than I'd thought before. He's more like the cliché caveman. His son hangs behind, pale and gauche.

When I wrench open my eyelids, Tom is whittling a stick or drawing designs in his notebook. Sometimes those designs are the ones I've just seen, and you can make of that what you will. For a moment I can see my rotting feet through my boots, the worms gliding under the earth floor of the shelter, my dad disapproving gently of the squalor round the fireplace, a caribou nosing a carrier bag, a raven with a human finger in its beak and a tree bowing gravely with a sweep of its arms.

Nothing will be the same again. I cannot believe in *mere* stones, or *mere* trees. Nor can I fail to believe in me or in Tom. The wood is suddenly full of story, of actors, of souls auditioning for a part.

In the spring the sun and the shoots and the flow of juice will tell their own stories. But now, in the winter wood, the onus is on me. The story has to be told. Something bad will happen if it is not, and there is no one else to tell it.

I remember that in many cultures stories can be told only in the winter or in the dark. Perhaps that's because to tell them at other times is to usurp the storytelling prerogative of those other more authoritative storytellers. I remember too that, just as it is disastrous not to tell a story, stories can heal, restore and redeem. Sometimes they can even raise the dead.

So, stutteringly and slurringly, I start to tell Tom a story. 'More,' he says. 'More.'

★

Soon he'll be a man. If I were a proper father, I'd tell the social workers to do their worst, and send him out into the cold to fast and fear and learn his own story. Just as, if I were a proper son, I'd hang my dad round my neck. Tom, though he knows that stories matter, doesn't yet know the right ones.

This is my fault, and I feel wretched. Modern rituals of initiation involve getting pissed on crap cider in a car park, losing your virginity in a bus shelter, being presented with a smartphone as if it's a Torah scroll and doing work experience in a chicken factory or, if you're middle-class, with an actuary. I suppose it's at least honest to start kids off the way they're likely to go on. Perhaps it would be unkind to equip a kid for ecstasy and woolly rhino hunting and then send him to a call centre for life.

The magpie listens, even in the dark. She sits a foot from my shoulder as I talk. When I'm talking, she's quiet. When I stop, she chatters for a while, and then nods at us and flies off until the morning.

★

I'm cold now. When I stand up and walk off for a piss, it takes some time for the blood to squeeze into my legs. I imagine my arteries as crushed straws.

The shimmering in front of my eyes is making me tired, because I feel I should concentrate on every little modulation in the texture. I'd like to eat, not for the sake of eating (which seems like a gross thing to do) but to make things humdrum again. I don't want each moment to contain so much. I don't want so much possibility. All the valency is exhausting. I don't really want to be free. I want a small menu of congenial choices, with not too much turning on my decision. To live with infinite possibility, and to bear bravely the responsibility for choosing between them: that's being a human.

Though I'm getting colder, the wood is getting warmer. There are holes in the snow now; the roadkill is starting to smell, and so are we. The cold cauterised our nostrils: now they are coming alive. I have a sore on my back that's the only hot part of me, and when it weeps it smells of straw in a dirty pig house – which I quite like. My breath smells of pear drops, even to me. That's the ketones of malnutrition. I'll fit into those old trousers when I'm back.

'When I'm back.' That thought is a bad mistake, and it can't be undone. It's a betrayal of the wood, and the magpie doesn't come any more.

*

We pack up. It takes a couple of minutes. We strip the tarp from the tree, roll it up and stuff it in our rucksacks along with the sleeping bags, the flints, the spears, the fire-hardened sticks and the notebooks.

X and the boy, clearer now, are standing by the barn. The boy has the golden eyes of a hare.

★

It will be hard and dangerous to go back. When you've been free and significant and lived as you're meant to, it's difficult to play the post-Palaeolithic charades. Which is why all governments (all of them spawned in the Neolithic) are terrified of people doing what Tom and I are doing. They hate, fear and envy the wanderers: the label-less, the self-possessed, the free. Just look at their legislation. They know that, once tasted, freedom (even if it's unwanted) will never be forgotten; that their lies will be clear; that their carefully constructed theme parks – which they call 'real life' – will be outed as fraudulent and fragile. No one who's been in the woods ever plays the game again.

All the same, I'm terrified of forgetting what happened here. What happened is that for a few moments we were part of this place. The hunter–gatherers *couldn't* forget that they were part of this. If they had, they'd have died. To be alive was to breathe, and to breathe was consciously to take the forest into yourself. Cetacean breathing is under voluntary control. Half of a whale's brain has to stay awake to tell the diaphragm to keep working. If a whale fell fully asleep, it would suffocate. It was like that in the Upper Palaeolithic. Stop taking in the forest and you died. It's the same today, if we only knew it.

Can I be a hunter–gatherer in suburban Oxford? Can I be part of every place I tread – even if my feet have shiny shoes? It's a matter of paying attention, I suppose. I can get better at that, just as you can learn, with years of practice, to notice that you're breathing. But now there's a systematic war on sustained attention and, like most people I know, I'm a casualty.

On the way from the wood to the train I snap a branch, and reflexively say 'Sorry, forgive me,' and find and eat an old blackberry and instinctively say 'Thank you'. I've made some progress. Gratitude is the main defining characteristic of hunter–gatherer communities. It's a gratitude rather different from the gratitude of the harvest festival.

*

By the time we've walked the mile to the village, we've moved 40,000 years, and so my eyes are roving, looking for entertainment. I'm modern again, and modern humans aren't entertained by significance, and so I've stopped looking for significance. Yet a medieval stone head in a wall winks as we walk past, and the magpie swoops down, perches on a spade and tick-tocks.

'I love that wood,' says Tom. I bite my tongue, and don't ask him what he means for fear of smashing up something precious.

There are flags on some of the houses in the village. They make me angry, and I'm encouraged by the anger. It's another sign that something has happened. I'm angry because I'm patriotic. Anyone who really knows and loves a place knows how inadequate a flag is. To suggest that a flag can begin to represent a valley or a tree is a blasphemy. No hunter–gatherer has a flag. Real patriots burn flags.

By the time we're at the little country station, moments start to process wearisomely after one another. There weren't any straight lines at all in the wood, but now they're all around: window frames, the corners of buildings, orderly lines of moments and purposes.

'It's all boxes here,' says Tom.

At Derby I eat a bag of chips, and the shimmering stops.

When we walk into the house, everyone gasps.

They're gasping not at X and the boy, who are standing outside in their skins, between the bins and a skip full of chairs, but at us.

*

Despite the twenty-four-hour minotaur roar of the bloody ring road, the gabbling of screens and the screams of warring children, the silence and eventlessness of Oxford are appalling. We'd been used to waves of wind breaking on the drystone walls, to the fizz of beeches, the honk and rasp of ravens and the rush of mice. Out there in the wood, everything changed all the time. Every time we looked up at the oak tree, its branches related to each other as they never had before and never would again. No shear in the top of a cloud had ever looked like *that*; no conversation between sparrows had ever sounded like *that*. That raindrop, spreading through the limestone, made first a fist, then a deer's head, then a chalice.

I struggle to see the same things here in Oxford. I agree with David Abram that 'there are only ever relatively un-wild places'. I bore people rigid telling them that even the most antiseptic shopping mall is part of the wild, crawling with contingency and fungi and exhilarating filth. But when you've had for a while what passes these days for real wilderness, the only-relatively-un-wild parts chafe painfully. I can't see the sky from where I'm sitting. A BMW goes past. Then a Ford. Then a DHL van, delivering ink cartridges to the journalist next door. Then another BMW. 'I can't talk now, I'm afraid,' says a friend on the phone. 'I'm really busy.'

No you're not. Your eyes and your brain haven't done in the last week what they've had have to do in ten minutes in our Upper Palaeolithic wood. And your arms, your legs, your ears, your nose and your touch receptors haven't done anything at all for years.

We think of wilderness as an absence of sound, movement and event. We rent out rural cottages 'for a bit of peace and quiet'. That shows how switched off we are. A country walk should be a deafening, threatening, frantic, exhausting cacophony.

If today's shorn, burned, poisoned apology for wilderness should do that to us, just think what the real wild, if it still existed, would do. It'd be like taking an industrial cocktail of speed, heroin and LSD and dancing through a club that's playing the Mozart Requiem to the thud of the Grateful Dead, expecting every moment to have your belly unzipped by a cave bear.

X and his son, in our bit of south Oxford, must think they're in a perverted kind of desert. The herds have gone. The birds are terribly silent. But even here X will have seen the skeins of geese overhead, noted where they graze and plotted his approach. Waterfowl were important in the cold weather. In some places they were life itself. X's balls might well be incubating snugly in a nest of swan down; he probably has duckskin gaiters, a parka of goose hide with the feathers lying on his chest and back, and a razorbill cap given to him by his grandmother; and perhaps his swan-foot fire-pouch is waterproofed with oil from a redshank's preening glands. I'm sure that his wind-scalded nose is glistening with coot grease, and I suspect that under his cap, next to his scalp, he keeps the stash of smoked duck strips that would keep him alive for a month if he didn't manage to trap anything else.

To get an idea of waterfowl in the Upper Palaeolithic, there's no point going to the Wash, or the Solway, or even to Iceland. You'd be better off sitting with your eyes shut on a bench in St James's Park, filtering out Big Ben, planes and the belch of buses, feeling the obscenely obese ducks bustle and fart around you. Or, rather better, drive as we did the day after we got back from Derbyshire, to Slimbridge, on the Severn estuary, remembering to have the car windows down, the heating off and wearing a thin T-shirt, shorts and flip-flops, and then walk, goose-pimply, to the main goose ponds, past the poison frogs. There you'll find goose and duck densities as in the old days, but it'll be tricky to get your harpoon past Reception.

From the hides bordering the river you can see Roman moorings and, if the sea has sucked hard, shadows in the far mud of something Mesolithic; the footprints of a couple of wolves walking side by side; some lumbering aurochs, some high-stepping cranes and some humans. The pale sky is sometimes dark with flashing lapwings, or silver with dunlin who turn faster than thought. There, for a moment, if you ignore the laminated guides to wing flashes and head shapes and the windows that hinge up so that you can poke your spotting scope phallically out into the wild, you can know something of the sizzling metropolis of the Upper Palaeolithic world; a febrile hum that has nothing to do with the hissing static inside human heads, or the scheming of a corporation.

'Why are they flying so much?' asks Jonny, our eight-year-old.

'Because they *can!*'

'Or because it is sunny, maybe?'

'Yes!'

When we get back, the winter really starts.

★

Our suburban winter wasn't so different from a winter in the wood – or so I tried to tell myself. We held on. We dug in. We wrapped up and rushed out into the cold for food and ideas. My ordinary life is foraging for scraps of ideas to make a meal: the only real difference between my sort of foraging and Upper Palaeolithic foraging is that there are as many or more ideas hanging on the trees or out in the woods in winter as there are in the summer.

We tried to tell stories: of the swan-roads in the air from

Greenland; of the light that is always cocooned in the dark; of the busy-ness under the snow; of the fire deep under our feet, in the core of the earth; of the Mind in the robin's head, and which extends well beyond it into our own.

We lay on marshes at dawn, our clothes frozen to the mud, listening to the geese come in, thinking of the noise their lovely long femurs would make if we drilled holes in them and blew, and wondering if samphire would go with their livers. When we found dead birds on the road, we boiled their bones and wrenched out their tongues with pliers, and saw that a blackbird's tongue is the shape of a stiletto. We saw that our local fox is nervous of plastic, saw that our local jay has a toe missing and saw the panic on the faces of illustrious academics when you mention the word 'feel'.

But most of all we watched: watched the stride of Orion from east to west, the stand-off between Taurus and Gemini, the creeping light leaking into the edges of the night.

X and his son vanished. We didn't see them after we left them on the doorstep. I wondered if they'd decamped into the tangled wood behind the nursery school, but I've never seen them there, though I walk and sit there at all times of the day and night. Perhaps they don't want to contribute to my story at this storytelling time of the year – because if a story is not your own, it's not a good story. Mind you, if a story's *just* your own, it's a really, really dull one, and so I'd have thought they might have helped out.

We were just playing at winter. The Derbyshire wood was always there, leaning sardonically in the corner of every room or sprawled on a chair; always interrogating me, often rudely:

'Have you thought what you might have been if you'd stayed there?'

'See that swelling paunch of yours? You'd feed a wolf pack for a month.'

'You missed something very strange in the life of that hare as you snored in the warm. How do you feel about that?'

'The frozen skin of the water at the bottom field formed a face. Whose, d'you think?'

The wood and the savannah are the past; the places where I and everyone were knitted slowly together. They are where our bones were forged, and they alone know what they're made of.

I cycled off to libraries and spent weeks searching for the recipes for human bones: Darwin, Dennett, Blake, Jefferies, endless anonymous scholars with pots and skulls and bags of flint on their office shelves and careful footnotes in their articles.

I held Tom's version of an Upper Palaeolithic hand-axe as I wrote and slept, hoping and half-believing it would be a super-conductive connector between me and the old dark; that it would help me to know how my bones were made.

Sitting on the ground in the nursery school wood, cursing the ring road, I took to turning my head quickly, knowing that the really significant things – such as the dead, or the past, or X and his son or the recipes for human bones – are always at the far frontiers of vision but can be taken unawares if you're fast or lucky enough, and if they choose to be seen.

The real work of connecting the past to the present was going on under my feet. Worms were pulling last autumn's leaves down into the earth, when the frost let them, and pulling up bits of Saxon farmers, Norman tax collectors and Victorian governesses – all to be carried down on the soles of the kids' Nikes when they head back from forest school, and wiped on the mat. Under the earth too, shoots were filling, uncoiling, sharpening and pushing eyelessly but with complete confidence towards the light that was coming, like ancient humans, from Africa.

We had plenty to be getting on with. We had old tunes,

deadlines, parties, waddles round the park, fumbled, flatulent attempts to understand one another and the loneliness that comes with failure, drunken hilarity and even greater sober hilarity, some unsettling moments from Njál's Saga and the invasion saga that is Christmas, and the feeling that in drinking farm cider you're onto something big and old – as if it's made of the wizened fruit of Yggdrasil (and if that doesn't make *Pseuds Corner* I've failed, and so has *Private Eye*) – and visits to graves in the hope that something would happen, dashes into surf, and crates of cheap red wine from the Peloponnese, and rushed dawns to catch the geese as they came over our house at exactly the same time from sunrise every day, and the hot fluttering bodies of the voles that we trapped in the back garden to see who shared the place with us.

We were marking time. X and his son must have been making in the dark the stories they needed to keep them alive for the rest of the year.

SPRING

'When spirits come in the forest something happens first. It gets quiet. You get about ten minutes of acute, padded stillness. It's not like any other kind of stillness, any other kind of quiet, any other kind of atmosphere. This is your moment to run, if you still have the legs underneath you.
Otherwise, the assumption is, you're in.'
Martin Shaw, *Small Gods*

In the winter humans wondered. In the spring they wandered. Though I fear the winter, I am lazy and sentimental and like stories, and it is hard to get me out of the cave, away from the fire, and to start being a story rather than telling one. It is easier to live on fat.

In the Upper Palaeolithic the spring was the lean time, when the ribs of the people showed, and the ribs of the land showed too as the snow retreated.

I have been hollowed out by the winter, though you'd never know it to look at me.

'You've got to get properly to grips with this shamanism business, you know,' a wise and pitiless friend had said. 'That *was* the Upper Palaeolithic world. To try to write about the Upper Palaeolithic without being an experienced shamanic voyager is like trying to write about swimming when you've never had wet feet.'

That overstated things, but not much. So I gritted my teeth, asked around and found myself banging on the door of a caravan in a Somerset wood.

'Come in,' called Polly, who smelt of sandalwood, 'and don't be frightened.'

Polly wasn't frightened of anything. She'd stopped breathing during a routine operation twenty years ago, walked along a tunnel, hugged her dead grandma and then been dragged reluctantly back by an energetic anaesthetist. She packed in her NHS filing job and went to central Asia, where she'd lived on horse meat, lichen and weekly doses of the fly agaric mushroom, *Amanita muscaria*. Sweating, twitching and nauseated, she held Judy Garland's gigantic meaty hand (with claws) and they stared together into the void before and beyond time, eating chocolate buttons and geckos. Then she'd worked cleaning the toilets at a wolf sanctuary somewhere along the Silk Road and come home to the West Country.

'None of that was as scary as the filing job,' she said.

'You've no reason to trust me,' she went on. 'Though you can. There are plenty of charlatans out there. People who do a weekend course in shamanism and then set themselves up in the business, with web sites full of photos of jaguars and oak trees and stars. We'll take it slowly. If you're ever frightened, you're in the wrong place, and we'll get out of it straight away.'

We did take it slowly, but even so it was all too fast for me. For weeks I went to Polly's place, lying on a foam mat next to the wood-burning stove, learning to shrink myself down so that I could go through a hole on a Greek beach; forcing my shoulders through the fringe of coarse grass that surrounded it, wriggling under the root of a old palm tree, stooping, then standing, always going down, ignoring the eyes in the wall, through a simmering stream, into a glade. There a fox waited for me, droning like a cat; yellow eyes with black specks in the pattern of galaxies I vaguely recognised. He *sounded* musky, for down here you could hear smells, and smell colours.

Sometimes he'd breathe on me, and it was like the wind on a hot spring meadow; sometimes he'd tear open my chest with his soft paws, and rummage around inside; and sometimes, if I asked him, he'd come back up the tunnel with me, out onto the mat and then in the car, all the way up the M5.

'It's good that he comes,' said Polly. 'Perhaps you're ready for Derbyshire?'

I wasn't at all sure, but Tom and I went anyway.

★

We get there just before the sun gives up. Our old shelter has weathered the winter well. A fox has left a twirl of dung right where I last laid my head, and I think that's auspicious. We sling the tarp over the branches, make a fire ('No: *wake* a fire,' insists Tom. 'Don't you remember?'), climb some of the old trees just to say hello and go off in search of the animals we know.

We know that the cantankerous old buck rabbit is dead, and there's no sign of the one-eyed robin, but the tick-tocking magpie is soon back on her old branch. Over the winter she's learned to hang upside down with one foot, and is proud of it. She jerks her head towards us to check that we're watching and approving.

The hare? I don't know. I can't see the silky tops of her ears in the hollow, but that doesn't mean she's not there. There's a full moon tonight, and if she's alive she'll be wallowing in it salaciously.

Tom's making everything neat; tidying up the stone hearth, gathering bracken to thatch the eaves where the tarp doesn't reach, beating down the undergrowth on the path to the clearing where he leaves bits of food.

'Let's not get too comfortable,' I tell him. 'We'll be shifting camp a lot.'

★

The main visible difference between modern humans and Upper Palaeolithic humans isn't clothes, or hairiness, or even our own physical weediness. It's their cosmopolitanism and motion and our own parochialism and sedentariness. Outside the winter they travelled widely, intimately and variously, their wits fizzing as they negotiated new treaties with the spirits of each new place and tried not to starve. We usually travel between identical places: identical boxes with identical facilities and identical food. And we sit, slouch or lie, with calories dropped by serfs into our open mouths.

In the spring and autumn the caribou made journeys hundreds of miles long, travelling up to thirty-five miles a day, hunters sometimes attached to them like remoras to a shark. We're going where I think the caribou might have gone. We'll identify the places where it would have been easiest to ambush them: a narrow pass, for instance. It's near such butchery bottlenecks that the seasonal conglomerations of Upper Palaeolithic people happened and that (later than X) the great masterpieces of the Upper Palaeolithic – the most sophisticated two-dimensional art until Byzantium – were created. Art, community, politics and religion all turned on the seasonal migrations of the caribou.

I don't like the idea of all that travelling. I feel I need to know how to know one place properly to become part of *its* story, and so to have the glimmerings of a story myself. The best stories are not just *from* a place, says the mythologist and storyteller Martin Shaw, but *of* a place. 'Find the place that has claimed you.' Well, there were moments in the winter when I began to feel that our valley was putting in a bid for me. Surely, I told myself, that was how it must have felt in the Upper Palaeolithic, when the boundary between the individual human and the non-human world was somewhere between the leaky and the non-existent, and the great governing rule was reciprocity: I take, you give: I give, you take; I claim, you claim. Shouldn't I stay in the valley and consolidate my claim and my claiming?

Well, I can't. They walked, and so must I.

★

She's still here! Not spreadeagled in the moonlight but tripping chastely around it: slipping from one pool of dark into another. Her belly is big. It's swinging from side to side, and she's bracing herself for the swing like a sailor on a heaving ship.

The hollow smells of fox. No wonder she's abandoned it. There are probably four foetuses inside her, and she can't just think only of herself.

But the hare and the magpie are scant consolation. The wood itself has turned away.

★

Sometimes, in the day, we skulk away from our patch, jumping behind trees or flattening ourselves on the ground at the flash of a distant walker's coat; watching them with wolf-eyes as they pass, drooling at the thought of the chocolate in their backpack; wondering what damage it must do to them to walk for hours through these old hills and then go back to the car park and turn on the news. Wouldn't they be better staying at home and not putting themselves through a change like that? Sometimes, with a snuff-movie appetite, we lie by the big road, thrilled and appalled by its violence and unconcern and recklessness and wailing. Sometimes, in the dark, we see inside some of the village houses: children feeding and squabbling; couples cooing, warring and ignoring. It's very rare to see anyone look at anyone else for more than a moment. We try to work out their stories; how everything fits into the stories they've chosen. Why *that* wallpaper? Why that print of a dying toreador? Why fry fish tonight, of all nights?

★

There's killing and the avoiding of killing all around us in the night. A barn owl combs the field as systematically as the scanning arm of a photocopier, dropping to a squeal once in a while. Foxes play tug of war with a ripping rabbit. A tawny owl misses a vole and flops into a bed of nettles. There's a badger in the distance, spitting and snarling as it is dragged out of the earth with tongs and dropped into a bag (we have no phone to call the police). And no doubt in the village a man is beating his wife.

Tom's asleep, dreaming of chocolate and reindeer. I'm awake, wondering where X has gone and whether his absence, and indeed the whole retreat from me of the wood itself, is because of some unrepented cruelty or casualness. That at least is an authentically Upper Palaeolithic thought. Did I ignore someone who needed to talk? Did I eat that roast pork carelessly, not thinking that it was from the buttocks of a near cousin who was still out there, watching every mouthful? Did I laugh at someone's misfortune? Any of those things would have shut and bolted the doors to the wood, and sent X away, revolted. Jay Griffiths talks about 'wild kindness': the kindness marking those who get their morality from the ground. That doesn't mean that teeth and claws are the source of kindness. I've been listening to those howls and screams, and don't want Darwin as my moral tutor. Competition, death, waste and dislocation don't make us good. The wood seems to expect more of humans than it does of owls.

I rake through the past couple of months, turning over old stupidities, jokes and posturings, and suddenly, as the church clock strikes three, I have it! A sneering, self-aggrandising question at a lecture. I think it can be dealt with, but it'll need something radical. A fast would be ideal, but I want to get things right with the wood before we leave it tomorrow morning, and so I'm up and out, stripped naked in the wind, walking along Tom's track to the stone of the rotting food. And there I stand with my forehead against a tree, teeth chattering some sort of inchoate *Miserere* until the next hour strikes and I can head back to the shelter. As I settle back into my bag I see the moon making Tom's track sparkle and hear X's soft hide boots behind the wall.

★

'Dad,' says Tom in the morning, 'Do you know why the grass on paths shines at night more than the stuff on each side? It's because grass *reflects*. It's all the silicon in it. Moonlight bounces off the silicon crystals.'

Yes, Tom, yes.

★

We head west up the dale, wading through the dew, with a sack of dried beef, a groundsheet and a shared intuition that this is where the caribou must have run. It's cold: the hungry sort of cold that sucks all the will and the joy out of you. If there were caribou just ahead, their sour smell (like a cellar where everyone's been eating sauerkraut and belching competitively) would drop to the ground with their dung, and soon even the dung would stop steaming and give in to the cold. The cold means that we'd know the location of a big herd from a way off, for we'd see the cloud of steam from their breath. Upper Palaeolithic hunters followed a column of cloud just as the Israelites did.

Today I am watching the deep purposefulness of crows. They have an intention far more complete than my own fractured intention. That crow there, flapping heavily from one field corner to the other, is far more of an *agent* than I've ever been. And agents know their own: the non-human world – of which Upper Palaeolithic humans were a part – has an integrated solidarity. Agency means meaning and significance. The field as a whole *means* something. It has a reason and a direction.

This mystical claptrap is one of the ruling axioms of hunter–gatherer life – as fundamental as the primacy of the market and the holiness of the profit motive are to us.

There are moments (and only moments) when I can align myself with the intention of the crows, and when that happens it's absurd not to believe that the movements of birds are more reliable auguries than satellite weather maps or a medical biochemistry lab.

The spirit fox is trotting somewhere by my side. Sometimes I see the grass sway as his tail brushes it, or bend as he presses it with his pads. He is steady, this one, but even so I can sense his excitement as he scents the caribou ahead.

We're climbing steadily, threading through gates and over stiles. Long ago the land folded along the seam we're walking and then flattened out, but not completely. There are more sheep skulls than sheep, and a brace of magpies is nailed to a barn door. A blue tit hops beneath them, eating the maggots the wind shakes out.

Tom is tracking the caribou. He finds it much easier than I do. He hasn't spoken for a couple of hours, just grunting to shut me down when I point something out or try to reflect. He's quite right. I'm only talking to cover my own failure to do this job properly.

The track crosses a road and leaves the seam, following a contour through a beech wood. Tom doesn't like this. 'This is wrong,' he says. 'They'd be down there,' and he speeds up to get this section done. To our right there's a long black ridge of smashed stone teeth. Beyond that is heather and some burned-out cars. The path ahead drops down to the river and walks alongside it for a while. Under the path there's a slow-running river, water crawling through the ground at a few inches a year. The rain that fell on me and rolled off my nose as a seven-year-old when we picnicked on the high moor to the north might be the same rain that's making my boots squelch now.

Tom is a long way ahead of me now, walking faster and faster, strangely bent over. Sometimes his fingers touch the grass. 'Slow down,' I shout, but he takes no notice. I would have stopped long ago. I'd like to sit by the river, eat a squirrel and think bucolic thoughts. But there's no chance. He's out of sight now, squeezing through hedges and (his footprints show) running some of the way. Perhaps he can't bear my company. I don't blame him. Or perhaps he wants to leave the road behind. I don't blame him for that either, though it's a gentle enough road, with just the odd Land Rover and trailer taking woolly black cows to market.

We can't avoid the market town, and I don't want to. My father bought red waistcoats with brass fox-head buttons here, and I came to finger partridges and teal in the game dealer's shop, and here I tried to buy an electric fence to keep my sister out of my bedroom.

Tom's waiting for me on the bridge. He's been trying to lasso the trout. 'But I'd never want to eat one. They live on chips and condemned meat.'

'Condemned meat'? That's an expression of my father's, but he never used it in front of Tom, and I never have.

'What's the hurry?' I ask, not expecting an answer, and

getting only: 'We should get on, that's all.' And off he goes, pushing north-west, bending, touching, through small old cold fields with white stone walls, rooks tumbling, a fox on the wall top wondering what we're bringing, a big killer-bird slumped in a treetop, full of flesh.

I think I know now where Tom is going, though I'm not sure he does. Seven miles ahead the path funnels into a slice through the hill. It's a slice notorious for murders. If you stamp on the grass the hollow bank gasps, and there's always a black bird somewhere, ticking off people who come through the gate, and another at the top of the gorge, checking them out. Most of the world here is underground, and even in the brash spring sun, and on the top of the ridge, the dark is in charge.

Tom is at the gate to the pass. 'Here,' he says, emphatically, and pushes through the gate and walks on up. There are small caves on both sides. He nods at them. 'Not there,' he says. 'Not yet,' and he makes for a rib of rock pushing out of the hill. He climbs to the top. 'This is it,' he says (not 'This will do'), and throws down his pack.

'Now what?' I ask. He shrugs, and settles down, watching the gate, where the road begins to climb steeply. There's not much to see. Day trippers from Sheffield and Manchester park up, wander fifty yards up the road and turn back. The odd emaciated cyclist, shiny and black, grinds past, eyes behind the glasses locked on a small square of tarmac near the front wheel. Bumblebees, rabbits and clouds browse. Tom's eyes don't move.

The day trippers drink up the last of their tea, reset the satnav and head home. The bees drag fat yellow bags of pollen into their tunnels, leaving the valley to the rabbits, the clouds, the black birds and us. Tom feels round for his bag, pulls it to him, undoes it, pulls out a down jacket, puts it on and pulls the hood over his head. His head doesn't move.

I take some dried meat from my bag and hand him a piece. He takes it silently. I make a fire in a rock cleft just below him, using wood we've carried a long way, expecting him to come down and warm his hands, but no.

There wasn't much sun here in the day, but what there was has followed the bees into the hill, and now there's a boat-moon rocking over Sheffield, sunk in the night. And still Tom sits. Part of me is put out. This is my project, not his. I'm the sensitive one, the one with the psychic pretensions. I'd like to think that this is paternal protectiveness – that I don't want him to be connected to skirling wildness, somewhere down in deep time, where I can't keep an eye on him – but probably I'm just envious. He's here, in this place, in a way that I'm not. I'm all over: part of me in Oxford, and in train time-tables, syllogisms, neuroses, hopes, medieval Iceland and the biology of spiders, and all of it ruled by memes hammered out in Greece, translated in Germany, misinterpreted by my dad, and rejected as soon as I fall asleep and start dreaming. If I could only be in one place, as Tom is, time travel would be trivial. If you're confident in one dimension, you can be confident in all.

I nod off, and wake because the fire has gone to sleep. Tom is still up there. He's motionless but taut and awake. I walk up and sit beside him. 'Anything?' He shrugs again. This time, shamed into wakefulness, I sit with him.

He's right. This is where they'd have come, and where men would have waited. This valley is the easiest staircase to the high lands to the north-west. A big herd would have scrambled through the bottleneck and spilled out onto the slopes. They'd have smelt men, lions and big cats as soon as they got into the valley, for the scent would have been trapped, just like the caribou. They'd have panicked. Some would have tried to turn back. Snorting and screaming, spraying them with fusty urine, taking our their eyes with scimitar-hooked hoofs, they'd have shoved the others who were still trying to come up and who hadn't yet smelt the death, climbed on their backs, taking them down, forcing their faces into the mud, snapping legs and popping eyes. And then, before the retreat could be consolidated, men would come out of the cave near the car park, shouting, and off Tom's platform, shouting, and flint spears would drive into the mass and the afternoon's grass would spill out all over from their bellies and the valley would fill up from the bottom with steam and blood and rumen liquor. Then the men would stop shouting and stop throwing and instead fall down and cry and do whatever passed in Upper Palaeolithic Derbyshire for the sign of the cross and spend the next few days skinning and roasting and scraping and trying to placate all the souls that were trotting round the valley wondering what had happened to their bodies and why their legs were drying in the sun rather than being attached, as is more usual, to each corner of their torso.

There would have been many men, women and children in and around the valley then. They would have streamed there from all directions, some for hundreds of miles, knowing that the caribou would be there then, calculating their time of arrival by the sun and the moon; knowing that there would be feasting, storytelling, religious ecstasy, sex in the woods and marriage-brokering; knowing that they'd see acquaintances from the wider tribe they hadn't seen for a year: people with whom it would be sensible to mix their DNA, with whom their daughters would do well, men with a better class of flint that could be swapped for hides, men with beaver skins to trade for women or red ochre.

We've got used to thinking of hunter–gatherers as egalitarian, and so they were, compared with any modern corporation or any modern society – at least for some of the year. But there are fashions in archaeology and anthropology, as there are with everything else, and anthropologist David Graeber and archaeologist David Wengrow have disinterred and rehabilitated some old work on the Inuit and the hunter–gatherers of the Pacific Midwest that chimes beautifully with what we see in the archaeological record of the Upper Palaeolithic.

In those more contemporary communities politics and sociology shift with the seasons. And the seasons themselves were determined not primarily by temperature, precipitation or light but by the movement of animals and the growth of plants. For much of the year the basic unit was as I've described it for X and his son: small foraging bands, their composition determined by family ties or function. There were root-diggers, deer-spearers, beaver-trappers, berry-pickers and, sometimes, hearth-tenders and fort-holders – for life wasn't always relentlessly mobile.

These small bands were relatively free of toxic presumptions about status. In this phase of their life everyone was obviously needed: their functions were interdependent, their contributions crucial; berries were chewed up together with the roast reindeer. No doubt there were boorish husbands who thought that hunting was better than gathering, but there were smart wives too, who rightly pointed out that there were many more plants in the family diet than there were animals. The general picture is of a fairly level playing field.

But in the big seasonal blood-fests it was different. There the (predominantly male) hunters were the main providers, and in the gathered clans there were many more male egos around, prodding one another to the diabolical synergy you see in any boardroom or cabinet. And with the greater numbers a stratified society, with rules and enforcement, emerged for the season and then dissolved until the next year. There were policemen with bison-hide boots and judges with mammoth-ivory ear-rings. Round the necks and wrists of the judges and the advocates jangled the dead, and the dead leaned on the counters at the police stations and made the speeches and drafted the judgements.

Upper Palaeolithic hunter–gatherers, then, drifted in and out of hierarchies, sampling a number of political and sociological possibilities from the seasonally changing smorgasbord. They were far more politically experienced and sophisticated than we politically monocultural people are. This will matter when we come to the seismic change that was the Neolithic and the great mutation that was the formation of states.

What forces determined the social and political architecture of the Upper Palaeolithic? Some say that Upper Palaeolithic hunter–gatherers were, and we are, simply Machiavellian chimps. Chimps are craven, vaunting, violent, submissive, manipulative and obsessed with status. Yes, the story goes, we have some cognitive software installed that chimps don't have, but that doesn't make for real differences in our nature. We are just upgraded chimps.

It doesn't work. Humans aren't – or at least weren't – *simply* anything. That software upgrade created a capacity for relationship (we'll come to that shortly) that had never been seen before, and *not just relationship with other humans.* Why should it be so restricted? A typical modern human has fifth-order intentionality· 'Peter believes (1st level) that Jane thinks (2nd) that Sally wants (3rd) Peter to suppose (4th) that Jane intends (5th) …' Just imagine what such a brain, but without any anthropocentric prejudices, would do if it found itself in a forest, buzzing and pulsing with otherness and agency. In fact don't imagine: try it. Talk or hypnotise yourself out of your biases, then catch a train to somewhere you feel uncomfortable, then sit for four days without the distraction of food or sex, knowing that you're being appraised by a million sets of eyes, some of them compound eyes, and by many other types of sensory organ, some of which probably know what's going on in your pancreas. Know that your appraisers are much older than you are; that they are much better than you at being in this place; that their minds, like yours, extend far beyond their skulls – perhaps, like yours, beyond the edge of the universe; that the electrons in their neurones are affecting the spin of electrons a billion miles away. Then let them in, and ever after check your mail for invitations from them.

I mustn't get over-excited. Upper Palaeolithic hunter–gatherers didn't always walk through woods as if they were on LSD. But they did take their ideas of self, and hence their ideas of relationship, society and politics, from the non-human world that bled into them.

That didn't mean singing 'Kumbayah' beatifically by forest streams, though it did mean saying, along with St Francis, 'Brother Ox', and it did mean knowing that the rain falls equally on the just and the unjust, and that everything is dependent on everything else, and that, as we are all always in the cross hairs of fatal contingency and the trigger finger is tense, none can boast.

But there wasn't enlightened, compassionate democracy on the tundra. There was status and, though it has been over-played by Darwinian orthodoxy (co-operation, community and altruism generate much of the complexity of the natural world), something of the free market too. Dogs really do eat dogs and, more importantly for the hunter–gatherer world, stags battle stags. We can't say that the Upper Palaeolithic world took its cue and its shape from the natural world without concluding that it took horns, and that sometimes those horns clashed. The horns had a purpose, unlike most modern horns, but they were still sharp. Evidence of human–human aggression is sparse in the Upper Palaeolithic, but hierarchy – even among the smaller hunting and gathering bands – seems likely. There were big stags, small stags, and there were hinds.

Horns are not the whole story, though they are more visible than other metaphorical human structures in the archaeological record and in the inferences we naturally draw from our thoughts about the structure of the world. Remember *My Big Fat Greek Wedding*? The husband might be the (notional) head, but the wife is the neck, and she can turn the head anywhere she likes. She's in control. Surely it was like that in the Upper Palaeolithic – as it is in all decent cultures. Everywhere, thank God, is really a matriarchy, despite the bellowing of the stags. William Irwin Thompson observed:

> Because we have separated humanity from nature, subject from object, values from analysis, knowledge from myth, and universities from the universe, it is enormously difficult for anyone but a poet or a mystic to understand what is going on in the holistic and mytho-poeic thought of Ice Age humanity. The very language we use to discuss the past speaks of tools, hunters, and *men*, when every statue and painting we discover cries out to us that this Ice Age humanity was a culture of art, the love of animals, and *women*.

Male power and status, like erections, are transient. Female influence, like menstruation, cycles eternally and must prevail. Only women produce live calves that can produce other live calves, and so on, for ever. Male hunters just bring home dead calves, to be roasted and eaten that week.

That's what I was thinking as I watched the council car park for ghostly caribou.

X isn't here. Why should he be? He's not my valet or tour guide. He's got work of his own and paths of his own. And perhaps he just doesn't like us.

I'm watching Tom, trying to make out what he's seeing. He gives nothing away. When the sun comes up over Sheffield, he finally turns away, rolls over on the rock, pulls his hood over his eyes and goes to sleep. It's unkind of him to leave me in my own head, which, like this valley, is a growling, haunted place. A bit of company would be nice. The days are harder than the nights. The nights are *supposed* to be haunted. That's the natural order of things. But mornings are supposed to be so full of hope, bacon rolls, lists, school runs and cancelled trains that there's no room for the ghouls that are prowling here. But biology finally exorcises them, and I sleep too. The last thing I remember thinking is that my dad got those pine needles for my exams from just over that ridge.

Tom kicks me awake. 'Time to go.'

'Time to go where?' I ask, and he gestures to the north.

'But why?'

'Come on.'

England's spine starts near here, running up through bogs, moors and high clean pastures to Scotland, stiffening and dividing the land, looking down to mills, lonely farms and lonelier housing estates. That's where we're going, it seems.

My footprints in the peat are my places. This – all the way east to Sheffield – is my place. This is where I learned about place, and so where I learned about exile. I have never met a Westerner who is not exiled – always from the Shangri-La of childhood, and usually from the place where they began, and to which they belong.

X isn't exiled. He's so attached to the wood that he's still there 35,000 years after he died. No doubt he's attached too, and for ever, to the pavement by our Oxford dustbins. Now *that's* aristocracy: to own and be owned by wherever you put your boot.

'Will you come *on!*'

'Yes, Tom, I'm coming.'

★

I loved this place. Still do. I fed on it just as Upper Palaeolithic hunter–gatherers did. Without it I'd have starved. More than that: I'd have been stillborn. Every night I walked out until the neon street lamps gave up, and climbed a hill I called Sinai – for it was there that I had meetings with things that dwelt in smoke and fire. On the summit of Sinai there was always the sweet air of the moor – heather, peat half-way to coal, grouse-breath and fox – and always something new, just beyond scent and thought. I wrote fast in a notebook with a biro in the dark. It was almost automatic writing, for this, like the Narnian wardrobe, was one of the Celtic 'Thin Places', only a breath or a fur coat away from another universe, where the usual rules of perception and composition don't operate. My contribution to the business was so obscure that I've wondered ever since if I have anything to do with the thought I call mine. In the spring and summer I squashed flowers in the notebook, and in the autumn and winter I squashed grass. I was probably trying to say that the words were written in juice that came out of the hill, like the colour of the petals or the stain of the leaves.

And then I smashed it all, odious little snob that I was. The local comprehensive wasn't good enough for me. Oh no. So down I went to town on the number 51 bus, marched into the library, took out the *Public and Preparatory Schools Yearbook* and started looking.

My parents didn't have much cash, and I'd need a

scholarship. I found a big one advertised and, unknown to my parents, applied, and waited by the letter box for weeks.

'Will you take me down for this exam?' I asked my father, handing him the envelope.

He opened it, read the letter and was more still than I thought anyone could be.

'Why do you want to do this?' he asked, very gently, and I shrugged, just as Tom does.

'If that's really what you want,' he said at last.

He shouldn't have said that. He should have known that words like 'I' and 'want' weren't half as weighty or meaningful as the thing that 'I' was starting. He should have tossed the envelope in the bin and dragged me up a mountain to get me out of myself. Instead, with a respect for my wholly notional autonomy that horrifies and touches me, he taught me where to put apostrophes, gave me some tips on essay-writing and drove me a couple of hundred miles south, where I floundered through some trigonometry, messed up my French pluperfect, put improbable phrases into the mouths of centurions in my self-taught Latin, correctly identified the tibia, described a sunset in emetically purple prose, belaboured a drum kit as if I were beating a carpet and played a minuet on the piano as if were wearing a powdered wig.

For some reason they gave me a scholarship. They must have been mad. My father said, sagely, 'Well, this will be an interesting journey.' My mother cried secretly and bought a case of sherry.

I made my peace with my Sheffield friends, rather hurt that it wasn't a pivotal event for them, went up to Sinai to ask for a blessing, told the valley that I was doing this for its benefit, packed my teddy bear in an old tin trunk and sat numb in the back of our old Volvo as my parents drove me down to start the autumn term.

It was a disaster. I'm only just beginning to realise how

much of a disaster. At the start I thought it was a disaster because instead of saying 'Top hole, old chap' and toasting muffins over the fire, they read porn under the bedclothes and told me to fuck off and die. There wasn't a straw boater or a glass of Pimm's for miles around, and to go rowing I had to get a bus by myself at weekends to an industrial estate. I bolted my letters from home, and those talismanic fir cones and leaves, inside my bedside locker, and walled up everything that was sacred behind a hastily erected wall inside myself that has only just begun to erode. I could do ordinary school life without drawing on the sacred: indeed it was morally essential not to let the heathens anywhere near the holy. I lived simultaneously in the sacred and the profane places – a feat of self-generated schizophrenia that has made me unsure ever since what I mean when I use personal pronouns. We *all* did it: no one who's been to a boarding-school is fit for relationship or public office. I wrote my diary in Greek letters, which I knew the philistines couldn't read. One day they broke into the locker and passed round the letters from my mother and father as they smoked in the dormitory at night. I did my best to break their noses, but they were too many for me.

Slowly it dawned on me what I'd done, although of course I couldn't admit it to my parents. The easiest thing for me to recognise was that I'd betrayed my friends and my place. No one on our road in Sheffield would have read my letters. No one who'd been anywhere near Sinai would have talked about sex as the philistines did. So my wretchedness crystallised into a sentimental northern nationalism – a nationalism that demonised the south and everything in it.

Every night, before going to bed, I stood tiptoe on the toilet seat to get a view of the road, and sent my love and loyalty on the lorries heading north, and breathed in when a lorry came south, because it might have some mud on it from somewhere near Sinai. 'The men of the north of England,' I'd silently intone, 'I saw them for a day. Their hearts are set on the waste fell, their skies are dark and grey. And from their castles one can see the mountains far away.' Our suburban house in Sheffield was hardly a castle, but from it, with the eye of faith, you could just about see what I thought were mountains, and my heart was certainly set on the waste fell. Re-pledged to Sinai, with my mind full again of dark grey skies, I unlocked the toilet door and went back into the smoke.

The relevance of all this to the Upper Palaeolithic is that all my morality and all my identity were *local*: they seeped into me out of the north country with the brown peat water; they were encoded in what the curlew sang; they could be deduced from the gait of the fox in the wood under Sinai. It's like that for hunter–gatherers, who are made and sustained, unlike most of us, by the place in which they happen to be. I had to lose my locality in order to find it and to know its importance: it had to be snatched away so that I could snatch it back and vow never again to let it go. That school did the crucial snatching.

It's been the brown water for me ever since, though I live in exile in Oxford. I've felt ever since that I have to make amends for the betrayal, and until quite recently was nervous about returning to Sheffield. I thought the moors and valleys would get their own back. I couldn't blame them. I imagined falling off one of the gritstone ledges, pushed by an avenging gust from Stanage Pole, or breaking an ankle up on Kinder and being pulled into the maw of a bog and fossilised, or just knocked down by a milk lorry from one of the farms near Tideswell where I'd worked as a child but which I'd abandoned for the bright southern lights. It would all have been completely just. The best I could hope for was time to put things right.

That, I think, is how hunter–gatherers move through a landscape. For me and for them the land is full of agency: it demands attention and a response, and has not only the power to affect our lives, but an active intention to do so. It is not just a scenic backdrop to human drama, a stage on which we can perform. It is the leading character.

In fact the land and its sub-agent Chris, with whom I'd collected moths as a kid, forgave. Sinai had me back. I sat up there again, decades later, and it deigned to speak. It was very gracious. But I've learned that I can't presume.

Recently I finished writing a book about the sea. I had been very nervous about it. The sea is very big, and I am very small. I thought it was hubristic to try, and so I set the book in a little dowdy port in a little backwater – scarcely the sea, really – hoping that Poseidon wouldn't take umbrage at my presumption and come to get me.

We were in North Devon when I finished the last bit of editing. I breathed a huge sigh of relief and then went off with a friend and Tom for a swim in the sea at a place I love. The sea was lively, but nothing out of the ordinary. We had our swim. The friend and Tom walked out. I was knee-deep in the water, walking out too, when I heard a roar behind me. A wave the height of a house was rearing above me. I've never seen anything like it before or since. It was travelling far too fast to run ahead of it, so I tried to dive through it. I had nothing like enough momentum. It picked me up and threw me about fifty yards. My leg hit a rock and was smashed. I was pulled back out to sea. I didn't lose consciousness, and another smaller wave spat me onto the shore. Some bystanders pulled me out.

Paramedics eventually got there, and filled me up with morphine. 'I've got to get a vein before he finally shuts down,' said one of them, and so I presumed I was dying. I hadn't thought it would be like this. My first thought was how the children would deal with it. My second was to wonder just what in that draft had made my presumption so intolerable.

A helicopter came, and a clever surgeon screwed me back together again, so my presumption, as it turned out, wasn't deadly. But that second thought was an authentically Upper Palaeolithic one, though I'd phrased it in a rather Greek way.

★

'*Please* come on.'

'Yes, Tom, I'm coming. Sorry, just thinking about something.'

'Can't you walk and think?'

'Well, sometimes. Yes, usually.'

'We do need to get on. We slept too long.'

'Too long for *what?*'

No answer. We've been walking now for a good seven hours.

But he's off, due north now, up the side of a stream that first ran with meltwater from the glacier that kept X's grandparents in France. He comes to the head of the valley. I'm sweating and cursing a quarter of a mile behind, and can see him stop and then cast around like a foxhound at the check, walking in a wide circle. He does it three times, pausing to peer hard ahead of him. On the third circuit he's away again, breaking decisively towards the only tree for miles around – a rowan on the skyline.

Surely he'll wait at the tree? But only for a moment, to look back at me. I can feel his frustration and urgency, but I'm too proud and puzzled to be cross. Away now, heading for the centre of a ridge. I'm surprised. I thought he'd move from feature to feature, but there's nothing distinctive about the point he's going for now.

He's speeded up again, and I'm falling further behind

with every step. I lose sight of him for a while and start to worry, but there he is again, running now up the rise, ducking behind a stand of bracken, slinking round the head of a grit-stone snake.

This time he waits. Not only that, but by the time I limp up, he's unrolled his sleeping mat and smoke is coiling up from a fire.

'Here for tonight, I suppose,' I ask.

'I think so. Is that OK?'

'Sure.' It's obviously nothing to do with me.

So we lie on our backs, chew the woody stems of cotton grass, listen to the skylarks climbing, twist ticks off our legs, watch a kestrel following the ultraviolet labyrinths made in the grass by vole urine and feel the ground greedily leaching the heat out of us. The light stays on longer than the heat, but it's suddenly snapped off, and the stars are there. I have the usual thoughts about stars. This isn't telling me much about hunting caribou. But then this is nothing to do with me.

In the night something is mumbling fast, as if there's a lot to say but that whatever is to be said doesn't go well into syntax. It may be a spring being forced through a stone throat deep under our heads, or a sleepless troll, or it may be in me, or it may be that all these things are the same.

I sleep on my back. South American shamans say that jaguars won't attack someone who sleeps with his face up, because in the face they recognise a being like themselves. Whatever is out there or down there or up there, it is a personality, and I hope it thinks I'm one too and spares me out of solidarity. When I wake, soggy with dew, Tom is on his haunches by the fire, toasting grass.

'As they moved,' he says, 'there was a sort of tinkling. Did you hear it?'

'As what moved?'

'The stars.'

No, I didn't hear it. But of course the Middle Ages did, and built a whole cosmology on the sound.

Tom stamps out the fire, swings up the bag and is off.

That's the way of it for the next few days: Tom in the distance, sniffing, his eyes puckered in concentration and against the sun and wind; saying nothing that doesn't matter; chewing meat; hiding if he sees a faraway walker; coaxing fires; sleeping fitfully; rising in the dark so as not to miss anything. Most of the starlight we've seen was already on its journey to this moor when X and his son were hunting mammoth here.

I am red-faced, and the sweat stings. My groin is raw. I'm worried about a couple of my tick bites – one of them in my armpit. Our dung is black and spindly, and Tom's taken to leaving it on hummocks like a fox.

I can't convince myself that the spirit fox is here: he lives in the twilight of my mind, and now my mind is full of sun. Perhaps he's trotting at the level of the Upper Palaeolithic land, many metres below us. The peat here – the peat I think of as the essence of the north country – is only about 12,000 years old. In X's day great herds – particularly of mammoth – grazed the steppe vegetation up here, leaving nothing for peat-making and depositing thousands of tonnes of succulent dung from which sprang the trees of the Mesolithic.

And suddenly it's over. It's mid-morning, and Tom's halfway up a bank. He stops, looks quickly around and then turns on his heel and starts walking fast back towards me.

'Let's go back.'

'Now? Why?'

Stupid questions. The only real question is how. We should really go on foot, of course, retracing our steps, but Tom is clear. It is over, and we need to get back to base as soon as we can.

We're a good seventy miles away, and there are only tiny roads at the foot of the fell. It will take a while.

'At least let's wash,' I suggest. 'We'll have to get lifts, and I wouldn't want to sit next to me just now.' So we walk back a couple of miles to a stream, strip off, lie in a stone basin with our heads under a waterfall, shake off like dogs, roll around on the grass, peel our fetid clothes back on and dribble with the water down to a lane.

There, because God is good and favours the Upper Palaeolithic, a tractor picks us up and dumps us in a village, and a buck-toothed vicar with no sense of smell drives us to another village, and a laconic asbestos-stripper takes us to yet another village, and there we're scooped up by a fell-running hard man with a thin beard who smells just as much as we do, and who feeds us with chips and tales of terrible pain, cheerfully borne, all the way to a bus stop.

By dusk we're back in the wood, loaded with things to eat if we can't kill or forage. Tom doesn't bother about the fire. He's asleep long before the white owl drifts across Sarah's top field.

★

The first of the small summer migrants are arriving. They are all males, come to stake their claim to a territory before the females arrive. They are just movements in the budding canopy, and songs that divide up the wood. Near us there is a thin, sickly song. If we could get near the bird, we might see that he was being drained by lice that hatched in the Congo, and that the toes that clutch that Derbyshire branch are still caked with mud from an oasis in Mali.

The eruption of birds into the spring wood must have seemed to hunter–gatherers to be all of a piece with the eruption of leaves from sticks and flowers from earth. Every spring I spend hours gazing into the sky, and yet I've never seen the first little gust of chiffchaffs being blown in from the south. In the winter they're not there, and then suddenly, one spring day, they are. Far and away the most intuitive explanation is that, just like the plants, they retreated underground in the autumn cold, and now the sun is luring them out. It's not surprising that I've never seen one hatching from the ground: how often do you see the very moment a daffodil rises into the light?

The ground under the feet of hunter–gatherers, then, is full of live but sleeping things. The winter earth snores and shudders, if you have ears and feet sensitive enough to know it. Although burials don't seem to be very common in the Upper Palaeolithic, there were some (and obviously very many more than we've found), and to bury a body in the earth must have been like laying it down in the middle of Piccadilly Circus at rush hour. There was traffic all around: risings, fallings, stirrings, snortings, sleepy changes of position, wing-spreadings, leg-stretchings. In many traditions the Earth – and earth – is a sort of battery. In the resurrection traditions we return to the earth, are recharged, remade and reborn. In Homer the subtle body of a slain warrior comes out through his nostrils and returns to the earth. Many meditators insist that their mind coalesces more effectively with Mind if they are sitting directly on the ground rather than on a mat in a meditation hall. Perhaps mats interfere with the flow of *Qi*. Industrial farmers pour bone meal and blood meal onto their fields so that corn can rise out of dead bodies to power living ones. 'He's gone to ground,' we say, meaning that he is hiding in a safe place, biding his time and will return.

X and his son went to ground here over the winter, eating smoked reindeer, dried salmon and the odd fistful of hazelnuts, pushing out the dark with fire and force of will; seeing the fire as a fragment of the vanished sun; bowing to the soul of a great dead shaman trapped in a crystal hanging round X's neck; feeling the fever of an old woman back home and descending to the underworld to look for her truant soul, rinse it out and bring it back; drumming with their fingertips on a squirrel-skin stretched on a goose-bone frame until their eyes fluttered in trance and their tongues fell out to taste, like snakes, the scent of the spirit-host summoned by the drum. When the feather of a grey goose fell onto the steppe from the leader of a skein, X wove it into his hair. The goose knew three realms – sky, earth and water – and travelled confidently between them, and so could help X negotiate the pathways between lower, middle and upper earth.

The coal tar soap is back.

Something is stirring around us. Grass moves when there is no wind. The magpie, unafraid of us, looks down from the top of a blackthorn and jitters.

*

It's hard to get Tom to talk. He grunts assent or dissent, or a reflexive cliché, but not much more. But when he doesn't know I can hear, he sings and whistles. One whistled tune is a strange, plangent thing, '*La li-li-li, li-li*', that I've never heard before. We're much more attuned than we were to one another's body language – or at least much more dependent on it, in the almost complete absence of spoken language.

And so, once again, Tom has taken me to where I'm meant to be. It's not that Upper Palaeolithic hunter–gatherers didn't have language: they did. But I doubt it determined the form and colour of their world as my language determines mine.

Archaeologists' search for behavioural modernity has been a search for symbolism. Words are the ultimate symbols: 'magpie' stands for that black-and-white bird, though the word, whether written down or in an audible form resulting from a series of complex laryngeal manoeuvres, is plainly something very different from the bird. The word stands proxy for several things that it plainly is not itself: magpies in general, the particular magpie referred to, the quality of magpie-ness and so on. To use a word demands that the minds of the word-user and the word-hearer meet. They must both realise that the other has a mind, and that minds can and do meet. All this is very complicated. To use even the simplest word in the simplest way means making lots of very complex psychological and philosophical assumptions. You need lots of hardware and some nicely tuned software for that. X has both. He and his ancestors have had them for a long time.

To understand how he perceives this Derbyshire wood we have to go to Africa and to some basic evolutionary biology.

★

Natural selection isn't telepathic. It has to see something to get to work on it. It can see the efficiency of legs and the strength of claws; it can see changes in brains if those changes *do* something; and it has a particularly keen eye for behaviour.

Brains can obviously have a profound effect on behaviour, but the real estate inside the skull is tremendously expensive. The brain comprises about 2 per cent of total body weight, but consumes about 20 per cent of its energy. Brain tissue needs about twenty times more energy per gram than muscle, and it has to work hard to justify that discrepancy. It does indeed work hard. What big primate brains work particularly hard at and for is *relationship*. A loner can get by with a small brain. But if you want lots of friends, you need a big brain. And the deeper and more complex the relationships, the bigger the brain you need. Monogamous animals have bigger brains than promiscuous ones.

In primates there's a neat, straight-line correlation between the size of the frontal lobes and the size of the group in which those primates are typically found. Our frontal lobe size suggests that we should be in groups of 150. That's become known as 'Dunbar's number', after Robin Dunbar, the Oxford evolutionary psychologist who has done much of the work in this area, and who forged the theory of the evolution of language that I'm going to describe here.

Once you know about Dunbar's number you'll see it everywhere. It is (roughly) the size of military companies across history and across the world, of Neolithic villages around 6500 BCE, of English villages recorded in the Domesday Book, of eighteenth-century English villages and Amish parishes, of the larger regional clusterings of hunter–gatherers, of Facebook friends and of really functional commercial groupings (such as Gore-Tex, which caps the number of employees in each factory at 200). Why? Because human communities work best if everyone knows everyone else sufficiently, and so can trust them. Reciprocity and trust are far more efficient societal and commercial lubricants than status or financial reward.

Primates are special, and human primates are especially special. Forty per cent of the total brain volume of the average mammal is neocortex (the modern higher-level processing part). That figure falls to around 10 per cent for shrew-like mammals, rises to 50 per cent for non-human primates and soars to 80 per cent for humans. We think and relate much more intensely than shrews. Or at least we *can*.

A big group size makes evolutionary sense for primates on the African plains where we came of age. More eyes mean a greater chance of seeing predators; more teeth mean a greater chance of seeing them off. Yet big group size is costly too. It's stressful, and stress has important biological effects

– particularly on females. It stops them ovulating. In primate groups the tenth-ranking females, who are more stressed than higher-status animals, are usually infertile. The answer to big, stressful group size, for humans as for baboons, is to gather round you a cadre of *real* friends: friends you know and who know you. But, as Dunbar's number demonstrates, you can't have an infinite number of reliable friends. Friendships need neurological processing power (and so the size of our brains limits the number of real friendships), and they also need time and work.

In non-human primate communities friendships are mainly pursued and cemented through grooming, which takes up far more time than is necessary for hygiene. It's about much more than lice. Evolution has been very cunning. It has made it very pleasurable to be groomed, for grooming triggers some neurones (C-afferent fibres) that respond only to the light stroking of hairy skin, and cause an outpouring of naturally occurring opiates (endorphins). Those endorphins make us feel relaxed, happy and close to those who are near us when we have the opiate high. It's a while since I had much hair, but I can understand why chimps like other chimps messing around with theirs. In Palestinian Arabic there's a word, *na'i'man*, used only or mainly to describe the peculiar feeling of well-being given by a haircut. That there is such a word shows the power of grooming-mediated endorphin release. Friends who are close enough to groom you can change your mood and your biology. If you're a female East African baboon, you're likely to have more surviving offspring if you've got plenty of grooming partners.

We see the power of opiates in unhappier contexts too. If your receptors are flooded with opiates, you'll want to be left alone: think about the self-imposed social isolation of heroin addicts after a fix. If your opiate receptors are chemically blocked, you'll want desperately to be groomed.

You'd expect the amount of time that primates spend grooming to be a function of group size: the bigger the group, the longer the grooming time. And up to a point it is. You can't spend all day grooming. Food has to be gathered, leopards have to be watched, gametes have to fuse and sleep must be snatched. It's just not feasible to groom for more than about a fifth of the day. What's to be done?

★

This is where it starts to get really interesting. Remember the human Dunbar number of 150? To maintain a group that size simply by grooming, we'd have to groom for about 43 per cent of our time, which would be deadly. Something else has to make up the shortfall, and other things have. We have developed a number of other endorphin-releasing, bond-forming strategies that don't involve touching. They are, says Robin Dunbar, laughter, wordless singing/dancing, language and ritual/religion/story.

We're better and more strenuous laughers than other primates (we exhale and inhale when we're laughing; they just exhale), and no doubt the endorphins of a laughing policeman flow more lavishly than those of a laughing bonobo. Evolution seems to think that laughter is very important. It is buried deep in our physiology and doesn't need to be learned. If you tickle a child born deaf and blind – one who has never heard or seen laughter – she'll laugh and smile. If you can laugh properly, it's a very efficient form of social grooming, because whereas physical grooming benefits only the one being groomed (although reciprocal altruism means that the groomer is likely ultimately to be groomed herself), laughter benefits not just the joke teller but also the listeners.

Laughter is a credible candidate for the first significant force that shifted human social behaviour (and hence group size) away from the social behaviour of our close primate cousins. We laughed our way out of chimp-hood and into proto-humanity. (Even today we regard laughter as fundamental to human thriving. Over half of lonely hearts ads specify 'GSOH'.) Once we were on our way, the other factors identified by Dunbar could come into play.

To see the origins of dancing, we probably have to go back at least to the origins of bipedalism (which enabled efficient fruit-picking and helped hominins keep cool in the African bush, because it reduced the surface area on which the sun could hammer). To be functionally bipedal you have to be well balanced and co-ordinated: to have the makings of a decent dancer. Human dancing is really just ornamented running. Try to imagine even the most elegant quadruped dancing: it doesn't really work. That said, it could be, as the archaeologist Steven Mithen suggests, that humans got the idea of moving in a way other than their business gait by mimicking the motion of animals, as modern Kalahari hunter–gatherers do. If a biped walks like a zebra, she's dancing.

Not only does rhythmic dancing – particularly with others – get the endorphins surging, but it can also (perhaps partly, at least, through the opiates) produce altered states of consciousness of the sort we've met already: dissociative states in which the 'I' and the body are forced apart. If these altered states of consciousness were indeed important in the ignition of subjectivity, the association of a sense of self with rhythm and music might be significant: it might imply that music can expound ourselves to ourselves in a particularly eloquent way. That's certainly my experience.

It's hard, and perhaps artificial, to seek to distinguish between the movements of a dance and the sounds that generally accompany and stimulate those movements, but it's worth noting that auditory driving (such as the drumming that often accompanies the dance of hunter–gatherers, or the noise of a disco, or the drumming of a shaman as she ushers an initiate into the underworld, or the chanting of monks) when it matches and synchronises the theta wave cycles in our brain, can produce altered states of consciousness. These altered states of consciousness are themselves pleasurable, and may generate an endorphin-like appetite for more. The hallucinogens known to the ancient shamanic world (such as henbane, mandrake, psilocybin mushrooms and deadly nightshade) seem to contain analogues of the naturally occurring neurotransmitters that are released when theta waves synchronise. Theta-wave synchronisation has effects on evolutionarily ancient vegetative brain stem functions, buried deep in the fish brain, far below the brash young cortex – functions such as breathing and heartbeat. It has been speculated that this theta-wave synchronisation might permit access to old, very fundamental information in the brain stem: information normally hidden in the subconscious. We might, through drumming, dancing, singing or henbane, learn – no, feel – what it was like when our great-grandparents were snapping at trilobites. Since most of what we are resides far below our consciousness (our conscious lives are really very boring and trivial) this sort of self-revelation can be life-changing now. If you know – really know – that half of you is a Cambrian anomalocarid, it'll change your online shopping profile and your devotion to daytime TV. There's no reason to suppose that it was any less dramatic when it was first experienced in the East African bush.

★

The endorphin-releasing effect of any activity is much stronger when others share that activity. This makes perfect evolutionary sense. If Dunbar's model is right, grooming (whether the removal of lice, the telling of a joke or stamping your feet on the ground in time to a drum) has been co-opted by natural selection for the purpose of bonding. And it is shrewd economics to maximise the number of potential bondees. But maybe there is something else going on: perhaps *active participation* in the human community is itself being selected. Players are rewarded: spectators who sideline themselves are sidelined by biology and then by the community.

Music-making with others is the most potent mood-enhancer I know: better than St John's wort, or climbing mountains, or running hard until the serotonin flows and the pain produces its own endorphin rush. I play whistles, flutes and a little Celtic harp in folk sessions in pubs, and the trumpet in a college jazz band, and it keeps the black dog at bay, snarling frustratedly at the door, as playing my instruments alone at home does not. The leader of the jazz band, a highly respected surgeon, says, tapping his nose confidentially: 'You know, Charles, this really is the best fun to be had outside the bedroom.' He knows about endorphins.

Much of what I know about community, and all of my politics, comes from playing old tunes in pubs. The boundaries between the players dissolve, and with them the animosities. No one who has played the songs of long-dead farmers holds to the atomistic, billiard-ball model of the self, or thinks that autonomy is the only principle that should inform our decisions. Who's ever met an autonomous man? If you have, you certainly won't want to have dinner with him. Would there be any entrants for a competition whose prize was a night out with Immanuel Kant?

Time and space start to behave differently in those sessions, and not just because of the beer. Reels, too fast to finger when you try them in your kitchen, slow infinitely until it takes an age to move from that F sharp to that G, and between notes you might well find yourself forking hay onto a wagon in a Dorset field.

Musical instruments appear in the record of early behavioural modernism in Europe. The oldest instruments we know are 36,000 years old – flutes made from the wing bones of swans, found in the Geissenklösterle cave in southern Germany. And then there's a big hoard of vulture-bone flutes from the foothills of the Pyrenees that's about 35,000 years old. X may well have one in his pack.

These very various sorts of grooming get people together. And until they got phones, humans used to speak to one another when they got together. Some people still do, though the capacity for it is atrophying fast.

★

And so we come to language – the ultimate groomer, wooer, bonder and, therefore, the ultimate scourge and divider.

We don't know how old language is. I have trawled exhaustingly through the debate, which turns on semantics ('What is language?') and on the view taken about the relationship of language to the evidence of symbolism in the archaeological record. Some say that language is 2 million years old, others around 50,000 years. It's likely that the very early prototypes added little to self perception or the perception of the world. For those sorts of effects you need fairly sophisticated language.

Language requires a vocal tract that allows precise breath control, a brain that can do the job (which will involve handling syntax) and something sufficiently worth talking about to justify the expense of re-engineering the brain and the vocal tract.

The necessary vocal tract changes have been around for a while. Bipedalism shifted the throat down, allowing for a longer and more versatile larynx. And hominin brains seem to have been primed for linguistic ability for a long time too. A useful marker of the necessary brain function is the FOXP2 gene, which is involved in the control of other genes crucial for language circuits in the brain. Many non-human species have the gene, and the modern human version (which differs from that of monkeys and African apes only by two amino acids) was shared by Neanderthals, and is accordingly likely to go back at least 400,000 years to the common ancestor we share with Neanderthals.

Neurologically and mechanically, then, Neanderthals could speak. I've no doubt they did. Even basic language would have been very useful as they hunted, collected nuts, made fire (which they knew how to control at least 200,000 years ago) and organised their family units. But eavesdropping on their conversations would not have been terribly interesting. It would have been like listening in to those phone conversations on trains about spreadsheets. For the overall impression one gets about the Neanderthals is one of cognitive inflexibility. They were very set in their ways. Those ways worked well for them for a long time (I doubt that we modern humans will last anything like as long as the Neanderthals did), but conservatism is always deadly – particularly in an era of climate change such as the Neanderthals faced.

The Neanderthals were hugely impressive. They had to be to survive and thrive for as long as they did in the hostile environment of Europe. They were magnificent naturalists, good parents, compassionate and skilled carers of the elderly, superb toolmakers (many modern expert flint knappers cannot emulate their ability to shear off Levallois flakes, in which a point is adroitly harvested and can be hafted immediately) and effective, if rather limited, hunters. In these domains, and in many others, they were brilliant performers. Taken as a whole, they had in their heads everything needed to triumph over the ice and the lack of ice. But here's the thing: you can't take as a whole what they had in their heads. Their minds seem to have been rigidly compartmentalised. The naturalist didn't talk to the nurse, or the parent to the nut collector. What keeps populations genetically healthy is promiscuous cross-fertilisation. What keeps brains effective and their owners alive in times of trouble is promiscuous intellectual cross-fertilisation between different domains of one's own brain, and between the brains of oneself and others. The Neanderthals had neither, and so they died out, victims not, probably, of homicidal *Homo sapiens* but of cognitive sclerosis.

Neurological Balkanisation, like all other Balkanisation, kills.

Neanderthal conversation was about what they were going to have for dinner – which was what they'd had for dinner for thousands of years.

But it's unfair to judge the Neanderthals, or any other pre-modern hominins, just by the words we presume they spoke. There's a lot more to life than language. Indeed I've just been denouncing language's ability to tell the truth about anything.

The archaeologist Stephen Mithen, putting together everything we can surmise about Neanderthal communication, concludes that they didn't have language in the sense that we understand it, but had a mode of communication that he calls 'Hmmmm' – an acronym for 'Holistic, manipulative, multi-modal, musical and mimetic'. They talked with their whole bodies; they were adept mimics and mime artists, clairvoyant anticipators of cues and agendas; and their vocabulary and grammar were musical.

This thesis has not been universally applauded. But even if Mithen is wrong about the Neanderthals, it's worth bearing with him. For he goes on to postulate that, although human language as we have it seems to have triumphed over whatever the Neanderthals had, the earlier ways of communicating have not been erased: they still persist in us, and can be revived. Indeed, those old ways might be better in some respects, and so we *should* try to revive them.

What would Neanderthals have heard in the damp forests of Europe? Well, says Mithen, since the natural world is more a musical than a linguistic place, a brain wired to work with Hmmmm might perceive it more accurately and intimately than ours. A brain for which holism is a virtue rather than a vice is likely to do a rather more satisfactory job of finding out what the world as a whole is like: Neanderthals would have heard, Mithen suggests, a 'panorama of sounds: the melodies and rhythms of nature, which have become muffled to the *Homo sapiens* ear by the evolution of language'.

It was the explosion into the mind of *words*, thinks Mithen, that demolished the walls between the mental compartments of pre-modern hominins, allowing their formidable brains to work as a whole: allowing concepts to slosh freely around, spawning other concepts.

I don't know if this is right. Nor does anyone. Did words create behavioural modernism – the capacity for rampant symbolisation – or were they created by it? Perhaps it doesn't matter much. By whatever means, language arrived. No one doubts its power to create and curate alliances, to formulate and test scenarios, to slice the slavering, hairy world into manageable chunks, to help melt ice, tame wolves, control fire, control humans and control, eventually, its users. It may have demanded a terrible price, but it may also have given us ourselves, and because when we possess ourselves we can give ourselves away, it may have allowed a new form of relationship.

It may, then, have enabled us to think ourselves into others' positions, or at least allowed us to upgrade the ability we had before we had language. This is the view of many mainstream archaeologists and anthropologists, led by Robin Dunbar and Clive Gamble.

The greater the degree of intentionality, the greater the imaginative capacity, and the more compelling the stories that can be told. Most modern humans have fifth-level intentionality, which is what you need to populate imaginary worlds with characters, and, so far as we know, only behaviourally modern humans have ever had it. To be a Shakespeare, though, you need to have sixth-order intentionality, and few of us do. You're more likely to be a Shakespeare if you're female, and you don't need sixth-order intentionality to enjoy a sixth-order production.

These sixth-order facilities made for world-class grooming. Most of the real talking in the Upper Palaeolithic world, as in ours, took place in the evening, when (as with us) most of the eating was done. They gathered round the fire. It was fairly dark. Body language wasn't very useful in the flickering firelight. Spoken language was far better. Sixth-order Upper Palaeolithic Shakespeares held their audiences spellbound, and natural selection, say Dunbar and Gamble, was quick with its rewards. The Shakespeares got status, meat, sex and progeny.

Theory of mind/intentionality itself gave an enhanced sense of self: personal pronouns were put in bright italics. Hunters looked at themselves, said to an animal, '*I* am killing *you*,' went home to ponder what that meant and, while they were pondering it, carved sculptures in mammoth ivory of fat women with huge breasts and resplendent vulvas. Then, at night, around the fire, the sixth-order myth-spinners began to tell stories that asked and answered questions such as 'Where am I from?' and 'Where is my dead father?'

I'm unhappy with this model. I don't see why you shouldn't have hundredth-order intentionality with Hmmmm. In fact

I'd have thought that the sort of intuition that's an essential element of Hmmmm is more likely to allow insight into the minds of others than the labyrinthine formulations of a linguistically based model of intentionality. Here is Dunbar's illustration of sixth-order intentionality: Shakespeare must *intend* that the audience *believes* that Iago *intends* that Othello *supposes* that Desdemona *loves* Cassio, who in fact *loves* Bianca. Put like that, it is very complex. It's easy to understand that you have to be a genius to erect a coherent written story on that foundation. Certainly anything more than sixth-order is unimaginable. But the appreciation of the story, as it emerges on the stage, won't, for most of us, involve anything like the rigour of the writing. We won't even analyse it in a similarly propositional way to the degree that our fifth-order intentionality allows. No, we'll bypass analysis altogether: we'll use our inchoate senses and our understanding of the way humans work to appraise, in great gulps of understanding, the *whole* thing. Hmmmm is alive and well, and doing most of the real work of mediating our understanding of the world.

Yes, language is the predominant medium for the exchange of facts, but facts are relatively inconsequential compared to the expression of emotion and the formation and expression of identity. Language works at the level of consciousness, and as I've already observed, almost nothing of what we are is determined by our consciousness or resides in it. We're almost all below the surface. Most of what I am and what determines my actions on a conscious level wells up from my unconscious. Hmmmm is the language of the unconscious. It is a far older and more fundamental language than our strings of words. It was a language spoken by premodern humans, and elements of it are spoken by modern non-humans too. Certainly it is the main language spoken by very young modern humans, and when we speak to them, we discover that we are fluent in it too.

Baby talk (more politely, 'Infant-Directed Speech', or IDS) is a universal language. We can all speak it; we do all speak it when we're talking to babies, and experienced mothers don't speak it much better than the most callow first-time babysitter. We're hard-wired for it. Our average pitch goes up, we use a wider range of pitches, we pause more, we elongate our vowels and make them clearer, we use shorter phrases and we repeat more. In short, it is more *musical* than our usual speech, and this musical speech engages babies' attention far better than mere language.

Singing to babies does an even better job. Lullabies (which are very similar in melody, rhythm and tempo across cultures) improve babies' mood and encourage sucking – and hence weight increase – in premature babies. (All human babies are, of course, born prematurely compared with non-humans. They have to be because of their enormous heads holding their enormous brains: they'd never be squeezed out if they were left to gestate for as long as their bodies should.) This has led Mithen to speculate that the musical aspects of IDS evolved specifically to deal with the unusual and prolonged helplessness of human babies.

Music speaks to us before words do, and more deeply than words do. The depth at which it works isn't just metaphorical. I can't be talked out of panic, but I can be sung out of it. It's not just me. Music has a clearly demonstrable effect on our breathing, our heart rate and our blood pressure. It's working deep down in the brain stem, affecting centres that were ancient long before the neocortex was plastered over the top of our brains. It's used for physical healing in many cultures, it reduces anaesthetic and analgesic doses, and music therapy is potent in mitigating many of the problems of patients with autism, OCD and ADHD.

We all know about music's power to affect our emotions, and our emotions *are* us, in a way that our considered thoughts are not. Emotions are primary. Cognition is parasitic on them. When I'm happy, I think more effectively. But emotional responses themselves do a lot of life's work: they're not always mediated through cognition. Most of my decision-making is intuitive, though I might give an *ex post facto* rationalisation of my reasoning that, in my less self-aware moments, I might actually believe. It's in the domain of emotion that I store the rough-and-ready solutions that experience has taught me will work in most circumstances. Most of the time, when faced with a problem, I reach into that domain, whip out one of the prefabricated heuristics and fire. Usually it works. Only very occasionally do I use my neocortex to craft a bespoke solution to a problem, and when I do, I very often find that I'd have been better off trusting my gut.

Mithen concludes from this that 'music has a developmental, if not evolutionary, priority over language', and that 'the neural networks for language are built upon or replicate those for music'. Hominins sang before they spoke. They Hmmmm-ed. And then language, in league with evolution, marched colonially into our heads and raised its flag.

This maps perfectly onto the relationship between language / cognition and music / emotion that I see clearly in my own mind. Language is arrogant and imperialistic. It purports to rule, and has the power to affect significantly the way I live my life and perceive the world, yet it does not have the concordance with reality that it claims to have. It shamelessly promotes cognition, which is more or less irrelevant in deciding who I am or how I feel or what I do.

Language wasn't our first language either as a species or as individuals. It's not our first language now – and if you met someone for whom language was the first language, you'd think them forbiddingly cold and dry, and wouldn't dream of inviting them to dinner. We're still most fluent, both as babies and children, in Hmmmm and in music. Music is a very accurate medium for the expression of emotion – for that which really matters. There is significant agreement between the emotion a composer intends to convey and the emotion invoked by the listeners. That agreement isn't any greater if the listeners are trained musicians: we're all naturals.

The distinction between language and music is not absolute, of course. IDS shows that. It is both language and music. I'll listen respectfully to anyone who argues that music governed by the sort of mathematical rules that underpin some language is the most powerful music of all – representing an orgasmic synergy between Apollo and Dionysus, or the right and left brain – but I'll not accept that that's a defence of language. Here is my conviction: language is most powerful and least abusive when it defers expressly to music. This in turn means deferring expressly to the natural world, for there is no language there of the sort we mean when we talk about language. The power of mathematical Bach is precisely that his mathematics are those governing the rolling of the ethereal spheres, the crenellated lacework of snowflakes and the forces that can be tolerated by an albatross's shoulder.

No doubt, early in our infancy as a species, we mimicked the sounds of the animals that were so important to us. The mimicked roar of a lion no doubt preceded and grounded the first designated word for 'lion'. All subsequent attempts to render 'lion' were less satisfactory: more self-referential. It was better when we merely mimicked. It was a sign of a proper humility, and it was more accurate.

There is a *real* (not arbitrary, not culturally manufactured) correspondence between the best words and the things they represent. Mithen notes that among the Huambisa of the Peruvian rainforest, one third of the names for the 206 kinds of birds they recognise are onomatopoeic, and he cites the work of Edward Sapir, who in the 1920s made up two nonsense words, *mil* and *mal*. Sapir told his experimental subjects that they were the names of tables, and they asked each subject which table was larger. Almost all of them said that it was *mal*. Vowels behave similarly in very different cultures. They're not arbitrary: they relate in some mysterious but consistent and fundamental way to qualities in the real world.

All of which is a very long way of saying why I'm going to listen more to Tom and to the wood. I'll try to use words dictated to me by the wood itself.

★

The legs of the chiffchaff (the Huambisa would approve of *that* name) are like grass stems. Its beak is the type of forceps you use for picking up the smallest nerve or blood vessel. Yet its throat is mighty. It sounds as if it's the size of a cave in which a family of bears could live.

This bird blew in last night, riding the wind as a fairground man rides the boards of a Waltzer, stiffening one wing and then another, as the man pushes up and down with his leg under his skinny jeans. Ten days ago it was huddling on the radio mast of a ship in Algiers harbour, bracing itself to squeeze between the layers of hot air that would take it to France. The sand in the wind rasped its eyes, so it flicked the membrane across them, whirred and clucked, pushed off, fell almost to the lifeboat, was picked up by the cook and thrown into the air, and this time it burrowed into the right, tight space, back and breast pressed by the laminae, and off it went, with little to do but balance and hold on over the sea. Once a wave reached up and almost snatched it, but wasn't quite fast enough, and then suddenly there were olives and oranges and caterpillars the size of carpenters' thumbs.

But it couldn't stay. Its chest felt wrong wherever there

were olives, and its head was jerked northward. It bounced up the western slopes of the Alps and over the chalk fields of Burgundy, had a narrow escape from a cat in an accountant's garden in a Paris suburb, hitched a ride for a while on a slow train through the flatlands, baulked at the grey cold of the Channel and slept in a filthy tree with starlings, and then took the sea with a thin, bruised bunch of little brown birds from somewhere in Africa. The tug was stronger now. It almost pulled its beak-sheath off. Soon it flopped into the tree above my head.

Well, possibly. There's a story for you. And that's what language can do. Probably it's all lies. You may have got a feeling about that bird as I told the story, but you'd have got a more accurate feeling, and without the lies, if I'd written a Migration Symphony instead.

But there are some true stories in this land, and some true stories about me and about Tom and about X and his son, and about the spirit fox, the birds, the stones, the grass, my dad and the coal tar soap, and some of them can be told in words. Perhaps the really big stories do need words, or need words if they are to be big and also specific.

That was certainly what humans eventually decided when they discovered myth and religion. That was very early on.

The background, though, was that X sang the tundra and the mammoths, and they sang him. The world, it has been said, longs for our appreciation and reciprocates generously. Part of that appreciation is in the form of aria: X and the natural world sang to and of one another, and their songs became religious.

★

We can learn these songs, if we learn to listen.

That's what I'm trying to do tonight.

There are noises all around. Some of them are discrete: whinings, hecklings, crashings. I'm pretty sure they are out there. But there's a constant purr too, and I can't tell whether that's inside or outside my head. It is the sort of purr that comes from a white thing.

Tom and I quarrelled, if you can do that without words – and of course you can. We're lying with our backs to one another. He's awake. I can feel his eyes swivelling. He thinks there's something there, but because we've fallen out he won't tell me what it is. It can't be anything very bad – perhaps just a hedgehog or a Roman soldier.

I get up for a pee. It's something to do. There's a heavy dew, and by the time I walk back it has softened up my bare feet for the edges of the grass, and I've a fine mesh of cuts.

I can't settle. The purr is still there, and now there's something else. There are many abandoned mine shafts under this hill. The purr might be from a white thing deep under us, though they don't usually come to the surface.

Tom's still awake: still swivelling. There's a new wind, and a new sound – though it's not from the wind. It's as if someone's painting a tree very systematically and very slowly with a big brush, sometimes dropping a glob of paint on fallen leaves. There's a light in the barn across the valley, where nothing and no one lives. It's very stupid to leave a candle burning there: it's full of hay bales.

Whatever's happening, most of it is happening beneath us. We're lying on a skin. A bone is sticking into me. A gut is working down there. Perhaps it uses the route of the tunnels the miners made.

Tom's afraid. I put my arm round him and he doesn't push it away. His breath settles and his shoulders sink. Soon his breath is keeping pace with the brush.

I can't bear the bone or the hide or the creak of the hill's diaphragm, and I get up and walk towards the glade of Tom's food offerings. As I walk there's a stiffening everywhere. The trees and the nettles stand to attention. A rabbit freezes with its neck hard and quivering. It's horrible to make a whole *place* do that. There is only one fluid thing, and it's in the middle of the clearing, looking straight at me. It drops its head and nips off flowers, and then looks back at me. It is a deer, and it shines in the moon. It sees that I am fat and slow and unarmed, and goes back to the flowers.

I feel the grass around my legs shaking. I look down and see that I'm shivering. It's time to go back and continue listening. I bow at the deer. It raises its head again, but does not bow. I walk along the shining path and wriggle into my bag.

The purr is a conversation. I doubt the voices are human. There are no words, but something's trying to say something important, and is being contradicted by lots of others. It doesn't sound as though it will end well.

It would be wrong to sleep. Whatever the merits of the argument, that many against one isn't right. At least I should stay around in case things get bad. This wasn't what I had in mind when I resolved to listen. I'm not supposed to be a referee. If I'm going to be a referee in this place, I'd expect a bit more respect from the bloody deer.

I'd never sleep anyway, even if I didn't have such an intrusive sense of duty. It's too interesting. I'm trying to work out the sense from the shape and pace of the sentences, and I think I'm getting somewhere. It's to do with a right to light – a version of the kind of argument you get in the county court when someone's built their extension too high. Something's climbing too high, leaning over too far, spreading too much.

★

Tom shakes me awake. I sit up and look round. It's before

dawn, but there's a dazzling moon. Cloud is coming out of the ground. It throbs like a heart. It has reached Tom's shoulders. The purr-speak is loud now. The thorn trees crackle. Blue sparks jump between the spines. Nothing has been resolved. The wood wants me to do something, but more urgently now.

'It's not my business,' I say under my breath.

'There's no need to shout,' says Tom.

And now he's packing, pushing his clothes into the mouth of a rucksack he can't see. The cloud is up to his chin, rising like the sea filling a cave. He's pulling on his boots, and knotting the laces blindly. He rips the tarp down and rams it after the clothes. He's fastening the straps of his rucksack, and I can tell that I'm doing the same.

The wood either wants us out or it wants us in, but whatever it wants it wants it badly.

If we just go downhill, we'll hit a track. Can't miss it. It's like a motorway compared with what we're used to. If we go right at the track, we'll hit the barn that we call the chapel, and then we'll be in the poisoned cow field, and then we're sorted.

The cloud is over his head now, but not over mine. I can see the trees, and I think I see the antlers of a stag. The top of the cloud is absolutely flat, like a table. I reach out and take Tom's hand. We've said nothing. We run. I can avoid the trees but can't see my feet, and I put my foot in a rabbit hole and go over, taking Tom with me. I'm lucky not to break an ankle, or worse.

We pick ourselves up and go more carefully. We're on the motorway. We can feel the mud, churned by hikers. Nearly there. Here's the stile. It's locked. It can't be. It doesn't have a lock. We climb over. That's the chapel. No it isn't. The chapel's gone. There's intent in this wood, thick and sour. It's getting into my mouth and nose. The chapel must be here somewhere. No 'musts' here, m'boy. Fog doesn't froth. Don't wipe it off your face, then.

I want out of here. Ah, breathes the hill: Why didn't you say?

And the trees and the cloud stop. We're out. There's a wall of cloud at the very edge of the wood, perfectly straight, perfectly sheer; as straight as the tabletop. And we're running and running and running, silently but for our footfall and the blood swishing in our necks, to the bus shelter. We don't care that it smells of vomit. In an hour we're in Matlock.

'*La li-li-li, li-li,*' whistles Tom.

★

'One for sorrow, Two for joy, Three for a girl, and Four for a boy', goes the old rhyme about magpies.

There's an English superstition that if you see a single magpie, you'll have bad luck. I've often been with people who, when they see a magpie, look sharply away, or cross themselves, or spit.

They misunderstand. If you look hard enough, there is always at least one other magpie. That's the way of things. They're never alone. If you look hard enough, you'll always have at least joy, if not progeny and an everlasting dynasty.

The rhyme tells the truth: if you don't look hard, look out.

★

We've gone west, which is where my family have always gone when we're scared or sad or have decided to die.

We're sitting in a cave on the side of a limestone ridge, looking out over the Severn Sea. There's no sign yet of X or his son, but we can't expect them to appear when we want them.

The spirit fox wouldn't get on the bus. I don't believe in him, and that no doubt offended him. The business with Polly was too easy. It didn't hurt, and I didn't really need anything from it.

A real descent to the underworld would have involved a shift to another of the many dimensions that mathematicians know are out there. Our brains are valves which prevent us from entering dimensions other than our familiar ones. When the valve slackens, perhaps we inhabit more of the reality that there is. Perhaps, then, we become more real. Perhaps it's then that I smell the coal tar. The coal tar in this cave is suffocating.

This cave was busy in the Pleistocene. They've dug up the bones of mammoths, woolly rhinoceros, wolf, brown bear, cave bear, fox, arctic fox, reindeer, horse, wildcat, cave lion and many, many hyena.

It's not ridiculous to think that X and his son knew this place. Just now, Tom and I climbed to the top of the ridge, and if it had been a clear day we might almost have seen the Gower peninsula, where in 1823 William Buckland, the Oxford geology professor, found the 'Red Lady of Paviland' – one of the oldest known human burials. The 'Red Lady' is no lady. Buckland had assumed that 'she' was female because the red-ochre-stained bones (actually those of a young man) were accompanied by ivory rods, perforated periwinkle shells (probably worn as beads sewn on the clothing) and an ivory pendant fashioned from a pathological growth in a mammoth's tusk. No English (or even Welsh) gentleman, went the reasoning, would dream of decorating himself like that. Buckland also got the date wildly wrong. He concluded that the burial was from the Roman period. In fact it is 33,000–34,000 years old.

That's a long time after X and his son were on the edge of the ice in Derbyshire. The climate was kinder to the Red Lady than to X, and behavioural modernity was deeply embedded in the humans of Europe by the time the Red Lady was alive. But the Gower, 34,000 years ago, was still a wild, dangerous and eccentric place to be. It wasn't for everyone. It would have attracted frontier-pushing entrepreneurs, and frontier-pushing tends to run in families. There weren't all that many families in the whole of Europe, and I'll bet that quite a few of X's genes are sitting in the Red Lady's bones, now in the Oxford University Museum of Natural History. I go to see her every fortnight or so. I can stand for an hour, staring into her glass case, willing her to speak, though she doesn't have a head. They must think I'm very odd.

Even if X and the Red Lady aren't closely related, there would have been oral gazetteers passed around the campfire, listing the places to stay, the places to hunt and the places to worship. This cave might well have been on the list – either as a place to avoid, because it was full of teeth, or a place of epiphany.

It's getting dark. Out on the sea a container ship is hauling cars to the docks at Avonmouth. Wales is switching on. The road below is at a standstill: someone has cut himself in half on the central reservation and is lying in a pool of blue flashing light. Up here it's an hour or so to badger time. The owl is thawing out and starting to blink. The woodpecker unwinds its tongue from around its head, sticks it into a hole, calls it a day and bounces off to the far valley, because birds have homes too.

A few feet into the cave the light is wholly overwhelmed. Few people today will ever meet darkness like this. It's a succulent dark. My hands are strange to me. They come at me out of the dark, like a bat. They are not fully mine, and not fully the dark's. We go further back. Now we can't see the entrance.

Tom's happy to sit in the dark. I'm still unnerved by what

happened in the Derbyshire wood, and think that I'll be scared by the dark. But I'm not. Derbyshire is a long way away. We'd got ourselves tangled up in some local business. We were caught in crossfire. It wasn't personal. And even if that Derbyshire hill could bother to lumber all the way down here, it would never find us in dark clotted as thick as this.

Tibetan monks meditate for weeks in complete darkness. The challenge of this is often said to be the psychological challenge of sensory deprivation. There is no such deprivation. Not only do all the non-visual senses (including many I've never suspected I have) come alive, but vision itself tries to show that it's still the king of the senses by giving a psychedelic show: sprays of sparks, a rolling kaleidoscope of interlocking coloured hexagons, faces elongated and green, and then, when it tires, scrolling back through its library of old images – anything to keep the brain from ceding jurisdiction to sound or smell. So there I am on a Dorset beach, eating ice cream with my parents, and at the same time curled up in a hole under the summit of a Scottish mountain, and sitting in a desert with my head in my hands, and running barechested across a Yorkshire moor with the grouse spurting up from under my feet, and drinking cider on a Somerset farm with swallows swooping so low over my head that I can feel the wind from their wings, and drinking red wine on a Greek island with a poet who lost all his fingernails to the colonels, and finding my pet rabbit dead and stiff, and watching the sea for a winter and wondering why there are no gulls.

Every sound made here lasts a long time. It's a reminder that every sound in fact lasts for ever, and that it's only our poor hearing that gives the false impression that sounds are temporary. Nothing that has been started, whether a sound or anything else, ever really finishes.

Mostly the sounds are of dripping water and whirring bats, but when you get your ear in there is something else.

As water bleeds through the rock, it scours it. Every drop, as it moves on down, carries dissolved rock. All the water is liquid rock. The scouring has its own sound – though it's not a sound that would make a spike, or even a mumble, on the screen of a recorder. It's the hum of eroding rock resonating with the eroding cells of my own body, and it brings a sort of macabre comfort. It's quite something to share solidarity with a limestone cliff, even when the solidarity is in the fact that we're both dissolving.

No doubt this isn't really a sound at all. No doubt it is a thought, or a usually overlooked sensation of the body. I do not know whether thoughts are different from sensations.

My fingers moult. I'd normally say that my wet arse was numb, but what's actually going on with it is much more interesting than that. Numbness isn't an absence of sensation but a space in which new categories of sensation can show their mettle.

We've been sitting in silence for a couple of hours. It is too interesting. I can't cope with all the interest. I want to get back to the neat everyday world of the visual. I reach into my pocket for a lighter. The pressure of my hand against my thigh is gross: almost painful. The noise of my skin against the cotton is like a car crash.

'What are you doing?' whispers Tom, and his voice and his breath are a hurricane.

'Nothing,' I say, and pull my hand away.

We sit for another hour. Then the car crash again, and the hurricane again, and I press the lighter and the heavens split open and St Michael and all his angels smash through the roof of the cave and we're screaming.

When we can look at the flame, it is a cruel, stark, uncompromising thing. But it tells us that we're sitting in a big round belly, almost high enough for me to stand, which narrows down at the far end to a gut through which, for a while, I could crawl, but in which I could never turn. Or, if you want a different metaphor, we're in a roomy uterus, big enough for many fat foetuses, with a fallopian tube running into it from deep inside the hill, and a birth canal spilling out through dog's mercury and rosebay willowherb onto the ridge.

We light candles and place them around the belly, and then we sit again, and start to watch.

On the wall we're facing there might be a tree, or perhaps a fish, or a bird's wing.

'That's not right,' says Tom. He moves three of the candles, and there it is, right in front of us, the front half of a colossal cow, the tips of its horns spearing the ceiling, its nostrils flaring hot.

We both edge forwards, reaching out to stroke its nose and, when we've done that without being gored, we move up to the broad shaggy neck and down to the foreleg, which ripples with tension and power.

This urge to be tactile bemuses me. It shouldn't. We instinctively reach over a gate to touch a horse's face, or over a candlelit dinner table to touch the face of the beloved. Children stretch and grab, and put their palms, with widespread fingers, into pools of mud. Our modern bias, though, is to link vision and cognition. 'I see,' we say, when we really mean 'I understand'. 'Seeing is believing.' We conceive of the whole creative process as visual: we call it *imag*ination. But when we're what we most fundamentally are – when we're children, or when we're relating to animals, or in love, or trembling in a Mendip cave – we overcome that bias, distrust our eyes and want the reassurance about the solidity of the connection of our solid bodies with the solid world that only touch can give.

One of the commonest motifs in Upper Palaeolithic cave 'art' is the human hand, outlined in red ochre. We've made them ourselves on the kitchen wall by blowing through drinking straws and hollow bird bones. They have been variously interpreted. Some say that they are the hands of girls who have just reached menarche, come to the cave to be initiated as women, and that the red ochre signifies menstrual blood. Some say they show that the cave wall was seen as the membrane separating this world from another – the other world being the world to which the shaman went in a trance – and that the handprints represent an aspirational pushing of the membrane, perhaps as an act of obeisance to the denizens of the world on the other side. But it is widely accepted that the handprints were made before – sometimes long before – the sophisticated paintings of animals that are often found in the same caves, and that they often cluster around parts of the rock that were later incorporated into the paintings (around a knobbly foot, for instance, or an eye). Palpation seems to have preceded visual representation, and may have determined it. The foot of the aurochs may have been found first, and the rest of the body discovered by groping up from it.

I'm stroking the cave wall, wondering what else is there. I'm noting that symbolism begets symbolism: a nose suggests a hip; a hip implies a hoof. Once you're locked into this synergy, it's hard to stop it building a whole world.

Tom moves the candles again. A stork. And again. An old man with an eagle's nose. And again. A pig. And again. A fox. And again. Another fox. And again. Yet another fox. And my father coughs from somewhere just inside the gut, and the coal tar overwhelms the smell of depth and guano.

You would only paint speared aurochs in a place like this if it was important, urgent and religious.

It was important and urgent because it was a response to human death, which, along with birth, was the great fact. It divided the world. In Michelle Paver's great saga of the Mesolithic, *Chronicles of Ancient Darkness*, her wolf distinguishes between 'Not-Breath' and the rest of animate creation. So do we. We're obsessed with death. Ancient humans were so obsessed that they incorporated it into every part of their thinking. We're so obsessed that we energetically banish it from all conversation.

It's not just a human obsession. Bereaved chimp mothers have been known to carry their dead, rotting babies around for nearly ten weeks. Imagine the smell in the central African heat, and so imagine the devotion. Chimps investigate the corpses of other chimps, trying to understand 'Not-Breath'. They smell the bodies, examine wounds, pull at the arms, stroke and hold hands, groom the dead, stare into their faces, try to open the mouth, drag the corpse over short distances and make calls rarely heard in other circumstances. Death seems special to them, and the dead have an unusual status – just as the ancestors do in almost every culture other than our own. Not everyone in the troop can pay their respects. The privilege of communing with the dead has to be earned. Dominant individuals guard the body and chase away lower-ranked individuals. There aren't many modern human fist-fights to decide who goes to a funeral.

The special status of the dead is seen in Neanderthals too. Regardless of Neanderthal belief about the afterlife (there are some very dubious suggestions of flowers in Neanderthal graves, which might imply an embryonic belief: I for one am not convinced), the Neanderthal dead are, as the archaeologist Paul Pettitt observes, persistently associated with specific locales. They need their own place. They're different. The Not-Breath are not like the breathing ones.

It's a small step from associating particular places (perhaps burial places or designated memorial sites) with the dead to believing that the dead persist. That's particularly so if you've already learned that mind and brain are not the same. Out of Body Experiences (OBEs) and Near Death Experiences (NDEs) can teach us that. Many modern people have had such experiences, and there's every reason to suppose that they were even more common in hunter–gatherer communities, where not only was there a shamanic culture that specialised in inducing OBEs by arduous trials or the ingestion of plant hallucinogens, but OBEs must have been a common by-product of dancing round a campfire, running hard after the caribou, or fasting. Air, if you treat it right, is a psychedelic gas.

When I hovered over my own body in the hospital casualty department, watching with more or less detached disinterest the nurse's central parting, I also saw all of my own head. I could see the skin that covered the skull. Inside my skull was all of my brain. No brain spilled out onto the pipe leading to the nitrous oxide cylinder, or ballooned out of my ears. It was all where it should be. I could see its boundaries. Yet all of 'Me' was surveying the brain's boundaries. My mind and my brain weren't the same.

It's generally agreed that quite early hominins tended to believe that minds could survive the death of bodies. That's not surprising, given what went on around the campfire and on the hunting trail. Hominins knew that brains weren't the full story, and therefore suspected that spilled brains weren't the end of the story. For them, the individual human performance simply changed venue, and perhaps gear. Death didn't diminish individuals. Quite the contrary. If they weren't penned inside their bodies, they could do more than they did before. They could become more truly themselves. Paul Pettitt observes:

> For most humans death is not conceived of as an abrupt end of the individual but a transformation from one state to another, one which usually results in an increase in the power of their agency as they 'transcend' the biological world.

Once that belief is in place, *theology* follows naturally. Pettit goes on to say that the natural tendency of hominins to read meaning into natural patterning, combined with their conviction that death isn't the end, made it natural to believe in gods, spirits and a supernatural explanation for the universe. Gods weren't creatures of corn, cultivation and Neolithic hierarchy: they were there in caves; they were the chiefs of the lands of Bigger Mind towards which the beloved dead, loaded with grave goods, processed once they had escaped from their skulls.

The religious revolution (or, if you must, the numinous revolution) was both a result and a cause of symbolising. 'What are the dead,' wrote Paul Pettitt, 'if they are not symbols? Symbols of lives once lived, of past attachments and the ultimate detachment, their accumulated social baggage and agency maintained through material arts, group memory and commemoration?' Well, yes. All true, I'm sure. But the smell of coal tar is getting stronger. If it's symbolic, it's burning my nose. Yes, I know that hysteria produces real physical debility.

One of the well-worn metaphors for the relationship between mind and brain is the radio. The brain, goes the metaphor, tunes into universal mind.

One of the best-known sceptics in the US is Michael Shermer. He's the editor of *Skeptic Magazine* and has devoted his life to exposing fraudulent, naïve and unexamined claims of supernatural happenings. He's an honest man, and his honesty compelled him to tell the story of how, on his wedding day, a long-dead radio burst into life to play to him and his bride, Jennifer, a tune that convinced her (she's as sceptical as he is, insists Shermer) that her grandfather was trying to communicate with her. 'We sat in stunned silence for minutes', writes Shermer. '"My grandfather is here with us," Jennifer said, tearfully. "I'm not alone."' She concluded that the music was her grandfather's 'gift of approval'. Shermer himself concludes that the experience 'rocked me back on my heels and shook my skepticism to its core'.

When my sister and I went together, several months after his death, into the room where my father died, a battery-less radio switched on. We took it as a declaration of presence: 'Here I am.' Maybe it was just a weird radio.

It's depressing that the modern dead have to use actual radios. If we had Upper Palaeolithic brains, they wouldn't have to. Living brains were more like radios then. Everyone who ever speared a woolly rhino was a radio, receiving the dead and broadcasting to them.

★

We've moved further west, to another cave. There's nothing votive about this one – except in the sense that everything humans do is necessarily votive.

This one, even at low tide, is only a few yards from the sea. At high tide we'd have to swim out of it, and before taking the children there I've had to take my prudent wife carefully through some intricate calculations to show that even at the highest spring tides there'll still be room for our noses above the water line.

It's at the foot of a long, steep hill that drops from the moor. Up there on the top, merlin kill meadow pipits, and skylarks rise straight out of the ground into the air as if pulled by a string in the hand of a puppeteering god, pouring continuous silver sound from their throats.

The cuckoos have just landed there. They're sitting on the high trees at the edge of the coombes, asking loudly who else managed the journey across the Sahara and the Mediterranean this year, and marking the meadow pipit nest sites.

In the wood below the moor – the wood that hangs as a

fringe over the top of our cave – buzzards mew, ravens *kraak*, foxes screech and plants screech too, as well as creaking, clucking and lisping, as they cry out to bees and one another. The track down from the road is slotted and puddled with the hoofmarks of red deer, come to escape the hounds that harry them across the moor from August to Easter. I feel their eyes on me as I lug our gear down the hill: big, brown eyes, just as the cartoonists draw them, but with flies clustered like warts at the corners.

You'd never stumble across our cave, even if you could be bothered to zigzag ankle-breakingly through the wood and onto the unfriendly beach with its rocks, selected over millennia to be the worst size for human feet to walk on. It's much further than you'd dream of going. But if you got there, you'd see a frog's mouth pouting out towards Wales, and inside, a platform where we make our fires, and further inside a nest of seaweed where we sleep, all curled up together, burping limpets and mussels at one another and listening to the creep and suck of the brown salt sea.

The first dawn light picks out a porpoise, or perhaps a dolphin, on the ceiling. It's throwing its tail up as if it is trying to dive away from the sun, and there's a suggestion of a fish in its jaws. The fish is swallowed within the hour, and the porpoise itself vanishes under the surface of the rock by mid-morning.

Our schedule is the oystercatchers' schedule. When the tide is high we wait, sitting on a boulder, looking out at the waves, wondering where each wave started: the thrust of a humpback's tail fin somewhere off Brazil? A lump of ice sliding into Baffin Bay, with seals glissading down? The propeller of a freighter carrying palm oil from Panama? When we've decided on the wave's route, we weave stories about what it has seen and heard on the way. A lovers' tiff between the freighter's mate and engineer, with an operatic stabbing by the lifeboat. Or the shooting of one of the Baffin seals, injured by the slide, by an Inuit hunter in a skin kayak. Or a big white bird that has followed the humpback around the Atlantic for twenty years, for no reason that, if it could speak, it could say.

When the tide pulls back, we get off our boulders, pick up our bags and walk along the sea floor, tearing mussels out of their beds, levering up limpets, ripping away slimy handfuls of gutweed that turn to mulch in your fist if you're not careful, lifting up stones and grabbing the little green shore crabs as they scuttle for cover, hacking off ribbons of sugar kelp with a flint knife and raking the sandy plains for brown shrimps with a very un-Palaeolithic net made out of Mary's tights. Then the sea pushes us back to the shore again, and it's time to browse for samphire, sea rocket (horrible), Alexanders (not too bad) and sea beet (splendid).

Then it's back to the cave. The porpoise has dived by now. We rustle and fuss and breathe the fire back to life. It loves stiff seaweed and an old cellar door, and blares a violent violet and earth-core orange with all the salts and metals of the ocean.

We open the shellfish by putting them on a hot rock, kill the crabs with a flint splinter, stun the shrimps with a stone, scoop out the shellfish with a mussel shell and load everything into a pot with sea water. This is wrong, of course. There were no pots in the Upper Palaeolithic. We've tried cooking in a deer skin slung on a tripod of beech branches. It works well, but now we're lazy.

Soon there's a maritime mess, bristling with legs and claws and antennae, green, gloopy, rubbery, fibrous, hairy and muddy. Eating it is a bit like biting into a rugby player's buttock sometime deep into the second half of the match.

Then we doze, prospect, swim, imitate the birds, watch the ships, scrape skins, plait nettle-fibre fish lines, whittle rabbit-bone hooks, curse the planes, plot the movements of the stones we've marked to see how the waves are shifting the shore, and then the sea will have gathered itself back into the channel again, and once more it's time to comb the rock pools.

At night the fire might summon a bird out of the wall: a cruel, hook-beaked bird with a forked tail, which flexes its wings if the flames are high enough. But usually, and unspectacularly, there are little, creeping, amorphous things, or parts of them: a leg on one side of a crack, a spindly ratty tail disappearing into a nodule, a nose quivering behind a screen of lichen.

I've brought a barrel of cider down here. That too is inauthentic. Apples came here with the Romans, and the first systematic beer-brewing was probably in the Neolithic. But beside the people of the Upper Palaeolithic I'm a novice at manipulating my states of consciousness, and it doesn't seem wrong to enlist a bit of help from that Mendip farm with the swallows. The other justification is the truth behind *in vino veritas*. Cider strips away the artifice and reduces the distance between me and the othernesses that I know are part of me.

So now I'm sitting on the shore with a tin mug in my hand, with Tom censoriously drinking water and scratching maps onto a piece of limestone. It's as dark as it ever gets here. I'm looking out to sea for something, listening to the thump of marine engines, wondering whether gulls ever sleep, trying to pick a strand of limpet foot out of my teeth, distracted by the flicker of bats.

There's a sound in the wood behind. A falling branch. I whip round, and there, for a moment, but long enough to nod at me companionably, is X, with his son at his shoulder. They are standing under a tree (do they ever sit or lie?), wearing bulky skin clothes and boots that reach almost to the knee. They have something big slung on their backs.

I get up quietly. I'm going to go and meet them. To stand up I have to turn away for an instant. When I turn back towards them they have gone, leaving a vapour trail of coal tar soap that leads up the hill towards the road, and a whistle hanging in the air, which I'd swear comes from the boy's chapped lips: '*La li-li-li, li-li.*'

★

The weeks roll on, curling into one another as the waves do when there's a stout south-westerly blowing up the channel. Most of my thoughts are foam. The pace of my life and my heart is the beat of the waves, and the wave is the metronome to all the music in my dreams. We are inside the sound and pump of this place as we never were inside the Derbyshire wood. 'The yelp of the gulls', I write in a stained notebook, 'is the sound of distilled loneliness.' Pretentious nonsense. There is no loneliness here.

Our faces are burned with wind and glare, our eyes narrower, our skin brittle with salt and the soil that's been washed down from Herefordshire.

I fear leaving. Fear it most terribly. There's a bus that will take us along the coast road, and it seems like death. Here on the shore, with the congenial dead walking with us in search of lobsters, and the cave wall now a riotous menagerie, we are happy, and we use words like 'I' and 'you' with more confidence and gentleness than we've ever done. There's no spirit fox. I don't seem to need him.

The bluebells are flat and faded. It's been six weeks since the hawthorn pumped the smell of sex (which is also, as a matter of chemistry, if not metaphysics, the smell of decaying animal flesh) into the deep Devon lanes. As we climb up the hill towards the road, the waves roll in my head for a while, long after my ears have been unable to hear them.

By the time we're at the lay-by, though, they've gone, and there's only the swish and growl of cars and the piping of the meadow pipits working to feed the cuckoo chicks.

SUMMER

'*Richard Lee calculated that a Bushman child will be carried a distance of 4,900 miles before he begins to walk on his own. Since, during this rhythmic phase, he will be forever naming the contents of his territory, it is impossible he will not become a poet.*'
Bruce Chatwin, *The Songlines*

This, I tend to think, is what the winter was for. *This* is why it was worth holding on in the dark. It is now that there's a duty to live the stories wrought in the fire and the cold.

My problem, as I've already admitted, is that I have no stories, or none that can do justice to the splendour of the high moor, or the deep coombe, or a single wave on the sand, or one rasp in a guillemot's throat, or the taste of sorrel – let alone the laugh of a child.

It seems that the Upper Palaeolithic solved this problem: found a way of being the guillemot, of inhabiting the wave. I'm left out in the cold, though the sun is blazing on the moor, though every wave is full of sand that should scour away my carapace when I dive in and though the children are laughing loudly and belabouring one another with sticks.

I've gone west again. Further west this time, where the beech hedges are higher than houses, the ravens are fat and lustrous on the bowels of sheep, salmon sidle (though you normally need legs to sidle) in the black pools; where if you breathe through your mouth at dusk you inhale a massive, intricate, insect ballet; where otters undulate through cow parsley and slip between the several skins of the river; where whirring packs of auks flounder up from the wave crests with twitching beakfuls of sand eels; where the cuckoo chicks are five times the weight of their foster-parents; where the gulls eat ice cream and their own young, and are always pristine, with cold blue eyes; where the red deer dip their heads so that their antlers don't catch on trees that were felled six hundred years ago, and run from the wolves hunted to extinction five hundred years ago.

I'm worried, though, that choosing to be here means, yet again, that I've betrayed the north, so I've brought with me, in a Tupperware box at the bottom of my backpack, those leaves and cones that my dad sent me all those years ago.

We started off the summer by going back to the Derbyshire wood. There were no walls of fog, and no palpitating hills. Just hot thorns, languid sheep and a neat cache of used hypodermic needles in our winter shelter. I understand why the needles were there. It was a good location for an opiate-based altered state of consciousness – as we, on rather different grounds, had decided.

The story-less-ness of the Derbyshire winter and spring had disillusioned me. Yes, the wood and the moor had been haunted by people, and people were stories, and stories grew like fungi out of the soil, but the people and the soil – and so the stories – would be inaudible to me unless I grew ears of my own. That wouldn't happen unless I had a story of my own: and that depended on me having an 'I' worth speaking of.

My father, and X and his son, were silent now. They never were in life. At best, now, there was a strangled, breathless, frustrated attempt by my dad, typically at midday, to say something. It would really, really have helped if I'd dried out his larynx in the sun and turned it into a medicine box, as I'm sure X did with his dad.

So we'd gone back to the spring sea cave for a while, and stood proudly on the shell midden we'd left. How energetic and successful future archaeologists would decide we were.

We spent a couple of days binding sticks together to make a crude raft, paddled with driftwood planks. It sank on its maiden voyage, of course; or rather, it became a sort of submarine, floating a couple of feet beneath the surface before becoming waterlogged and going to the bottom.

The first known seafarers did rather better. *Homo erectus* sailors somehow got from Lombok to Bali more than 850,000 years ago, and in the Middle Palaeolithic, around 60,000 years ago, rafts (presumably) crossed the Timor Sea to settle Australia.

'Not a fair comparison,' said Tom. 'They had bamboo.'

They also had many personal qualities that we don't have, such as an ability to weave sails from rushes, bind poles together with plant-fibre rope to form a deck as flat as a snooker table and to smell land a thousand miles away. But there was one supremely important quality – one far more important than knot-tying or bitumen-plastering. That was the ability to know when a place itself is nudging you out of it.

The sea-cave beach was doing that now, but it took a long time to feel it. Our raft sank not just because we were dismally incompetent boat-builders but because we were supposed to be somewhere else. We should never have come back here – or at least not for a year. We'd been too happy here in the spring. That was our lot. Any Palaeolithic person would have known that, and would have noticed that there wasn't even the slightest whiff of coal tar.

By thinking it could have been home, and trying to make it so, we would have ruined it. Humans always foul their own nests. A properly sensitive person knows that, and always moves on. A kind piece of land knows that, and always says, gently or sternly, as necessary: 'Time to go. Time for the next place.'

We go time after time, day after day, year after year, back to precisely the same places. Our feet fall on precisely the same place on the tube station steps as they did this time last year. We sit in the same seat, urinate in the same ceramic hole, type the same computer keys, turn the same handles, speak the same things into the same phones. The only changes, by and large, and most of the time, come from things we regret: the creep of disease and birthdays; the budding, sprouting, deepening and coarsening of our children. We've become disastrously able to deal with non-event. We seek it with all our hearts, minds and souls, and build elaborate financial and psychological structures to try to bring about the bringing about of nothing.

No hunter–gatherer ever goes to the same place more than once. Yes, there are the great seasonal cycles. Yes, the clan gathers in the autumn to club the caribou together and copulate, and in the winter to tell stories, and bands go to the river in the spring to spear salmon from the stones where they stood last year, and to the bushes in the autumn which regularly give them basketfuls of berries, but that's not at all the same as going through the same gate at Bank tube station for decades. No bush is the same today as it was yesterday. A gatherer would have harvested it yesterday: there would be no point in going today. The bison might go down the same valley each year, and the same rocks might generally be a good throwing point, but the wind has to be reassessed every second, and no bison ever in the whole history of the world has ever taken exactly the same path before, or ever will again, as the one you're trying to kill, just as the orientation of the leaves above my head now has never been the same before, and never will be the same again. That's why being outside with your senses on is so much more exhausting than being inside. Everything, including you (for your cells are different from the ones that composed you just a moment ago, to say nothing of your thoughts), is new every millisecond. To notice it fully is impossible; to try is exhilarating and sapping.

A place is not a static thing. A field is not a field is not a field. We knew that once. The notion of stasis came late in human thinking. A field, a stone, a rabbit, and everywhere you can put your feet, are *processes*. As Iain McGilchrist puts it: there *are* no things. That's what a life defined by wandering can teach you.

This does not entail an ascetic disdain for matter. Still less does it mean that it is meaningless to talk about the self, or about the endurance of the self beyond death. Here are two pieces of bad anthropology, both from Bruce Chatwin's splendid book *The Songlines*, in which, having started correctly, he grasps pugnaciously the wrong end of the stick:

> The Bushmen, who walk distances across the Kalahari, have no idea of the soul's survival in another world. 'When we die, we die', they say. 'The wind blows away our footprints, and that is the end of us.'

> Sluggish and sedentary peoples, such as the Ancient Egyptians – with their concept of an afterlife journey through the Field of Reeds – project onto the next world the journeys they failed to make in this one.

Even were the assertion about the Kalahari San an accurate summary of their beliefs about the afterlife (it is not), and even if the assertion about the reason for the colour and complexity of ancient Egyptian religion were correct (it may or may not be), it is certainly not accurate to state as a general principle that hunter–gatherers have no conception of the afterlife. Exactly the opposite. They do, and that conception informs every step and every breath. The afterlife overlaps with this one. The departed have never quite departed. Death, as we've seen already, increases their agency. They have a potency with matter that is greater than, though rather different in nature from, the ability of incarnate humans to affect the material world.

If you've got an 'I' – a consciousness, a mind, a self – that can persist beyond death, you've got an I that can walk around in a Derbyshire or Devon wood looking for bears or reflecting on the sort of thing that you are. You've got what I'm looking for.

I wonder if the sense of self and the sense of personal immortality arose at exactly the same time. If you believe sufficiently in the sort of self that one's relationship to others shows you to have, extinction is ludicrously implausible. It's suggestive, to say no more, that signs of belief in personal immortality appear at exactly the same time as other signs of symbolic understanding – other exuberant signs of 'I-ness'. Behavioural modernism at least coincides with the birth of the afterlife, even if the two are not causally related. Think back, too, to what Robin Dunbar and Clive Gamble posit about the key ingredient of behavioural modernism: a higher degree of intentionality / Theory of Mind, which facilitates relationship and makes it bloom. If we're not vibrantly, clairvoyantly relational, we're not modern humans. One of the relationships that defines us – and one that tests our intentionality most arduously – seems to be with the dead. More obviously in the Neolithic, to be sure, but also in the Palaeolithic.

Here's the manifesto. To be human: relate – and not just to living humans but to dead humans too, and to non-humans.

To be human: believe that you will endure. To be human: wander.

The sea-cave beach helped us with all three of those. Tom and I quietly grew closer to one another than at any other time. The words fell away, and other older and more eloquent modalities took over. Our heads sometimes merged. And then, of course, there was the coal tar soap and the apparition of X and his son. And then the beach kindly cold-shouldered us and pushed us on, because stasis is death.

So here we all are. I'm tramping the moor, sleeping in a beech-branch bivouac in a triangle of woodland next to a pool, living on rabbits, whortleberries and tickled trout, coming back once in a while to the cottage where the others are when I feel guilty about being a dreadful father, and when I need someone to search my back for ticks and feed me lasagne.

Often the family – all six of us – is a roving band, like so many of the modern hunter–gatherer bands who park by the picnic tables. Sometimes there will be a specific target: spear-leaved orache, silverweed or dog rose petals for salads, or owl pellets to dissect so that we can know what's running around at night, or skins from roadkill for a coat, or a raven's feather because it seems necessary. But each of us is naturally a specialist in exploring a particular ecological niche, and usually, on our wandering, that's where we'll be.

Despite our aching political correctness, the roles divide naturally, unfashionably, and immovably along traditional gender lines. Ten-year-old Rachel is a prolific gatherer – a diligent bush-stripper and leaf-nipper. Eight-year-old Jonny is a prober of holes and collector of bones. Thirteen-year-old Jamie has a vulture's instinct for carrion. Tom's a tireless generalist, ranging far from the main party, usually stooped, often crouching, looking at tiny things and horizons, making and testing weapons, croaking and whistling. It's '*La li-li-li, li-li'* these days, more often than not. My wife, Mary, thank God, is focused, like Rachel, on berries and salad. And I: I moon uselessly across the moor, never really there.

The sea is about hunting, and here we're all killers, jerking mackerel from one world to another, and then to the next, pitiless because they don't have eyelids to signal to us. The shore is eyelid-less, and so we do terrible things. We probe crevices with hooked sticks; we club, impale and crush. There's none of the usual instinctual reflection about killing. Released souls need to be assuaged, we know, but we seem to assume that a thing has to have eyelids to have a soul.

That's not to say we're cruel: we're really not. But a screaming neurone is a screaming neurone, however simple the brain that processes the scream and turns it into an experience.

All our sensibilities, depravities and philosophical convictions give way to hunger – or, perhaps more accurately, to the fire on the beach. That fish and that crab were plainly meant to be in the fire, and so any moral queasiness about their route there ebbs away. The hearth is the great reconciler of persons and arguments. No one at a fire looks at anything else but the fire.

The nights are big. The past is close. The valleys clutch tightly everything that has happened in them. X and his son are often here now. They are thawing out, and I'm beginning to know rather better what separates them and me.

There are two main things. The first is their vulnerability, and so their sense of dependence on the world and on themselves. For all my talk in the winter about contingencies, and about our constant teeter on the knife-edges of despair, disability and eternity, I can't really replicate X's constant vulnerability. There are wolves in my life, but not as many as in X's. There are uncertainties, but not about whether I'll starve. X has chosen to be vulnerable. Back home in France he wouldn't be: here on the ice he is. But, chosen or not, his vulnerability is real. I can starve myself, and tell you how it feels, but that won't tell you anything about how it feels not to have the option of not starving. I'm play-acting, but no amount of literary method-acting will let me play X's part properly. That's a disappointment, because so much turns on vulnerability. All our relationships, whether with lovers or mountains or pieces of flint, turn on our appreciation of how vulnerable we are.

So perhaps it's impossible to explore properly the second main difference between X and me: the nature of the self. That's a bit of a worry at this point in the book, for this is a travel book purporting to tell the story of precisely that exploration. Perhaps I can't even get out of the door.

But I'm not ready to give up just yet. Let's recapitulate. X is at a time in human evolution into which, for however long it had been gestating, a new type of self-perception and self-understanding had burst. It was manifested in a new symbolic sense. It may have been ignited by that sense, and no doubt was fanned by it. It wasn't the first time that consciousness had existed: far from it. But the new human consciousness may have been different in kind from anything that had gone before; or was so different in degree from anything that had previously existed that it looks like something different in nature; or (my preferred option) it was so much better at expressing itself that it looked different in kind or degree from anything that had existed before.

Yet, it seems sensible to suppose, this revolution did not have the effect of sweeping away overnight all the old ways of relating to the world and to oneself. Even today, 40,000 years or so later, those old ways have not retreated very far. They are still within grasp, even if you have to dive deep on a vision quest or a psychoanalyst's couch, or to try to communicate remotely with a dog, or to charm the birds down from the trees, or to read the unconscious thoughts of a sleeping child, a demented parent or a comatose patient.

This summer I'm trying to grasp the old ways by having a symbol-fast. The idea is that if I can put symbols aside I'll have direct, unmediated contact with the things outside my self. That was presumably what it was all like before the symbolism revolution, and what it was often like for a long time afterwards.

As with physical fasting, it helps to build up to it gently. A day without text. A week without human art. A morning without speech. Then, slowly, longer bans on reading and non-natural images, longer retreats into silence and extensions of my usual meditation practice, in which I watch and feel the breath ebbing and flowing, and watch passing thoughts with growing apathy. The apathy doesn't grow fast enough for me, and I take to picking up each thought between my finger and thumb, as I'd pick up dog shit in a plastic bag, and dumping it outside my skull. At first the thoughts see this eviction as a challenge and redouble their efforts at entry and despoilage, but after a while they begin to tire.

Anyone watching me would see a middle-aged man with a beard like a badger's arse, dressed in an old seaman's jumper, mud-stained jeans and a woolly hat, sitting cross-legged with his eyes shut on a rock covered with raven pellets, for all the world as if he were trying to hatch them.

But no one's likely to be watching me. The rock is almost hidden in a fortress of new bracken. When I open my eyes, I can see the sea. It is close enough to see white crests if there's wind, but it's too far away to hear. Red deer hinds often feed on the edge of the wood, which boils up like broccoli from the stream. There's nothing to bring them to the bracken, but they'd never scent me in here anyway. Behind me, below the crest of the hill, there's a small cluster of standing stones.

After a while they seem brassy and brash, and I resent them. It's like finding a shopping mall up here.

There's nothing at all original about what I'm trying to do. It is the age-old quest for contact with reality – contact unprocessed by language, priests, systems of thought, pictures, presumptions, templates, vanities, rules or institutions. It is the simplest idea. It is merely to do what all young children do all the time, until we ruin them. The measure of what we've done to our children is the supreme difficulty of achieving, as an adult, this direct experience of anything at all, even for a fraction of a second.

The best-known literature of this quest is the religious literature that repudiates religion and is sceptical about the value of literature: the mystical literature of every culture that has ever existed, and the Western Romantic literature which began as a reaction to the monopolies of established religion and has taken on a new life as a reaction to the monopolies of materialism. Much of this calls into question the utility of non-experiential modes of knowing – and particularly modes mediated by language. 'The Tao that can be spoken,' said Lao Tzu, 'is not the eternal Tao.' Apprehension, not comprehension, is the only true epistemology. We can know something only if our view of it is obscured by the Cloud of Unknowing. St Paul was converted not by becoming convinced by a set of propositions, but by being flung to the ground in an encounter. Most religions are at bottom (where the mystics are to be found) a call for a new type of epistemology, and a call to wake up.

The non-Western world has been less completely divorced from direct experience than the West, and has never systematically denigrated direct experience as the West, quakingly fearful of it, has done. It's easier to know what it was like to be a very early Upper Palaeolithic hunter–gatherer if you're sitting in a Shiva ashram in south India than it is if you're sitting in a Calvinistic church in Kentucky or – its cognitive equivalent – a Wall Street office.

It's possible, though very hard, for adults to have true experiential knowledge of something through their own efforts. It normally takes years of sitting in a meditation hall, learning to watch and to feel your own breath, and so to inhabit your own chest. Most of us will die without ever having experienced anything at all. Those in the West who *do* experience something will usually have had experience rammed down their throats in a painful but merciful epiphany: will have been dragged kicking and screaming out of the library, the workplace and their neat assemblage of comforting algorithms.

Here is Andrew Harvey describing how it can happen:

> I listened and studied hard but understood little; my mind had become hardened by years of too-scrupulous training in Oxford scepticism and irony. I made 'scholarly' notes in a series of black notebooks in tiny handwriting, as if I were still back in the Bodleian Library at Oxford writing a 'thesis' on 'religion': I laugh out loud when I read these notes now; they reek of fear and uneasy smugness.
>
> But then, thank God, my mind and heart were forever shattered open by a series of direct mystical experiences that forever altered my perception of the universe and compelled me to become a seeker.

Those experiences included a sensation of the moment of his own incarnation, an encounter on a beach with a beautiful androgynous being who turned out to be him, and an occasion, walking by the sea, when his mind was split apart 'like a coconut flung against a wall', and he saw the boats and the beach glittering with brilliant bright light, and the waves 'singing *om* as they crashed on and on'.

'*La li-li-li, li-li,*' whistles Tom.

Something like this may be necessary before we can understand the Upper Palaeolithic. I've done my own apprenticeship on plains of burning grit, in ditches, in tree houses, in boats rocking between hot islands, in inner-city squats, in jungle ashrams where the cicadas chant in time to the mantras and on hillsides where the loneliness was the knife that took off the top of my head as you slice open a boiled egg and the desolation the spoon that scooped out my mind and smeared it on the scree slopes for miles around.

I don't suggest that the humans of the Upper Palaeolithic wandered through the tundra, smiling beatifically, drunk on epiphany. This was the Stone Age, not the stoned age. But it is not fanciful to think that, although there was an 'I' and a 'You', Upper Palaeolithic identity was less discrete: mind, though personal, was more distributed; boundaries, though real, allowed for more diffusion; personhood was seen as a nexus, not as a point. And they were instinctive Darwinians, realising the cousinhood of the non-human world; acknowledging the ecological, cyclical nature of humans, with all that meant for human responsibility. Humans ate animals; animals ate humans; humans ate animals; and so on unto the ages of ages, with plants sometimes interposed in the cycle. We're all in it together. It's hard to say where humans stop and aurochs begin, or where I stop and you begin. This jumps out from everything we know about the Upper Palaeolithic mind, and although we should be cautious about drawing parallels between the Upper Palaeolithic and more modern hunter–gatherers (anthropology is *not* archaeology), it is echoed in the ontology of some modern indigenous communities.

Shamanism – one of the most prominent features of Upper Palaeolithic society – built conduits between personas and domains. It is true that not everyone was a shaman, and no doubt there was some high-priestly posturing by some of the shamans, but everyone was a beneficiary of shamanic engineering and journeying. The shamans opened windows to other worlds, and everyone breathed the air from beyond.

There are many voices to listen to as I sit on my rock. Many wise tutors. Someone counted 2,070 species in a 90-metre stretch of Devon hedge. That presumably doesn't include bacteria, protozoa or other microscopic organisms. That's 2,070 different dialects. For, yes, the plants have acoustic lives too. Some plants produce sweeter nectar in response to the sounds made by pollinators, using their flowers as ears to funnel the sounds deeper into them; roots travel towards acoustic vibrations made by water; bubbles in the xylem pop; and plants emit ultrasonic signals that may be picked up by animals and other plants. Those ultrasonic signals differ in stressed and unstressed plants: a computer can be taught to detect the difference.

We're not aware of being aware of these sounds, but then we're not aware of most of the influences on us. Only a tiny fraction of the information passing between humans is information we're conscious of receiving. Most is beamed in under our conscious radar in the form of body language, pheromones and possibly other ways that are dismissively lumped together as 'telepathy' or 'mere intuition'. We know beyond argument – it is everyone's experience, as well as being the subject of many systematic investigations – that being outside in a green space affects our mood. 'Forest bathing' is good for you. Is it really so psychotic to think that among the forces lifting our spirits might be the emollient and perhaps downright poetic whispers of plants? And if that might be right, is it ridiculous to think that an uncluttered human brain, less seduced by the powerful PR of the cognitive machine, might be even more affected by these and other voices? I don't know, of course, and it doesn't do to be too vacuously tree-hugging about these things, but why not?

I can't pretend that plants chatter away merrily to me. But I do hear the changing moods of the willow warblers, the ravens, the dogs, the farm labourers and the sea. For though the sea has always been inaudible to me here, I can hear it now. Perhaps a muscle in my inner ear, tensed to protect me against the fusillade of industrial sound that is the backdrop of life, has relaxed, letting subtler sounds through the door. But however it happened, there's a low lisp, a semitone or so lower than the night breeze in the beech above my bivouac; the lisp of rolling stones and sliding sand.

I mustn't give the impression that it's all contemplation out here. In fact it is terribly busy, and very cognitively demanding. I'm constantly having to figure out solutions to problems that, if I couldn't run back to the cottage, would be matters of life and death. How do I make that green water clean? Will those strips of rabbit meat dry, or just rot? Is that a storm coming in from Ilfracombe way? Should I build a side-wall to the shelter? Is it time to move to the next valley? Is that root poisonous? Are *all* ghosts friendly? If I hit the flint just above that bulb, will the whole thing shatter? Were the snakes in last night's dream caused by the lichen on the tree bark I was burning? Can the farm down below see the fire? Can I float all the way down to the sea without being seen? How many tiny brook trout is enough?

In our instinctive racism, we tend to think of hunter–gatherers as simple people. Not a bit of it. Hunter–gatherer life demands a far wider range of skills than ours. A typical modern orthopaedic surgeon might have a diet of 75 per cent hip replacements. And she'll have to dictate reports, counsel patients, do a bit of gentle management and drive to and from work each day. That would be a life of suffocating simplicity for a hunter–gatherer. Being a super-specialist is easy and dull. Being a generalist is hard and interesting.

This isn't noble-savagery, but it is ignoble-orthopaedic-surgeon-ness.

★

The symbol-fast is working, up to a point. Part of our legendary human plasticity is that we can forget quickly as well as learn quickly.

★

I can't maintain the fast when I'm back with the family. I've tried. It lasts for twenty minutes at the most. This isn't because there's pressure to read bedtime stories (though there joyously is), or because the bookshelves allure (though they do), or because I'm tempted to bark instructions (I'm not), but because here, in the home, are the only problems and dilemmas that matter, and the only way to deal with them that I really understand and trust is to *plan*: to fabricate hypothetical worlds and try out in them various hypothetical solutions until I alight on one that works. An hour ago I was living in a real wood. Now I'm living in an artificial world: one that certainly doesn't exist now, and probably won't exist ever. 'Give us this day our daily bread'? That's not good enough for me. At any rate I don't have the faith necessary to live that way. It's a faithlessness grounded in a laughable over-confidence in abstraction.

★

X is usually there whenever I go out on my own. I've never had a good look at his face, but I know he's dark-skinned. That shouldn't be a surprise, but it is. In fact the dark-skinned humans who crossed from Africa to Europe about 45,000 years ago remained dark-skinned in western Europe, DNA analysis tells us, until about 8,500 years ago. I wonder if he can make enough vitamin D in our anaemic light, and then remember that he's dead, and so I needn't worry.

This sets me wondering about what sort of entity he is. There are many possibilities, I suppose. They've been much discussed by the mystics over the millennia. Sanskrit, as you'd expect, has a particularly sophisticated taxonomy. As well as the food body, which is made of meat and eats food, there are energy bodies, astral bodies and infinite bodies. The non-meat bodies are multi-dimensional – perhaps in the way spoken of by mathematicians – and so one wouldn't expect them to be bounded by our conventional dimensions of space and time.

The dissolution of X's physical brain and the passage of 40,000 years only mean that I'm mad to see him in twenty-first-century Devon if there's a compelling reason to suppose that quantum phenomena apply only at the level of elementary particles, and not at all at the level of bodies composed of aggregations of those particles.

We can't know for sure whether the icy plains of the Upper Palaeolithic were stalked by humans with non-meat bodies. But we can say that different types of body were needed for shamans to operate effectively in worlds contiguous with this one. Typically the other bodies were those of non-human animals, and there are some suggestions (the part-passage of bodies through cave walls that's depicted in some of the cave art, and the absence of a baseline on which the animals ran, leading to a floating appearance) that bodies in those worlds didn't behave in quite the same way as in ours. It's really not far from Lascaux to the Vedas. The real costume of the Upper Palaeolithic is not a horse-skin cloak but a saffron robe.

★

Tom refuses point-blank to go on one of the initiation retreats that I'm keen for him to do.

My keenness isn't part of my experiment in hunter–gatherer living – although vision-quest-type rituals are very common in hunter–gatherer communities. It stems rather from a long-standing conviction that marker posts are important: that passage doesn't happen properly unless there are rites.

What I have in mind is four days fasting on Dartmoor, supervised at a distance by a wise, kind friend. That's it. And then we'd see what happened.

What's likely to happen is three days of increasingly painful scouring, as loneliness, fear and disorientation grind off the crud of presumption and illusion, followed, if he's lucky, by a day of far more painful filling, followed by a life-time of equilibrium, confidence and humility.

Traditional vision quests often involve an introduction to a spirit animal. That may or may not happen on Dartmoor. My friend certainly wouldn't suggest that it could – let alone that it should. What's more likely to happen is a creeping animism: a recognition of the aliveness of everything around – including, crucially, Tom himself.

Rituals of passage are unavoidable. I'd quite like Tom to have one that was meaningful. The usual coming-of-age rite is now, at best, the presentation of a new iPhone to replace the one they've had for years. Dartmoor would do better. Evangelical Christians talk of marriage as leaving (the family home) and cleaving (to the spouse). Technology does that very effectively. It takes the whole of the child's attention out of the home and binds it immovably to itself.

My own rite of separation was the wrench away from Sinai to that crass school. There was certainly a leaving, but

I didn't cleave to the thing or the place or the ethos to which I'd gone. Had I done so, I would have been destroyed, as happens to some degree to most of the children who are put through English boarding-schools. In the Faustian bargain that has been the mainspring of British institutional life for two centuries, I would have been lent, in return for my soul, the spurious authority and confidence so beloved of British voters.

Better to leave home for four days, cleave to oneself and the ensouled world of a Devon moor, and then either return or go – but in either case as an ontological aristocrat who'll hate putting on a tie ever after.

While caught up in these musings, lying alone by the fire, sucking a rabbit's leg, I notice that X has recently been alone too. His son hasn't been with him.

I panic, as if I've lost my own child. Perhaps I have. I scrabble through my memory. When did I last see the son? It was when Tom was last with me. Have I ever seen the son when Tom wasn't there too? I have not.

I don't know what to make of that.

'La li-h-li, li-li.'

*

I wake up one morning, the sun streaming into my face, the wood steaming all around me, and say loudly, with real distress: 'Pots!' I'm a very slow learner, and the penny has only just dropped.

The Upper Palaeolithic didn't have pottery. It wasn't that they didn't have the technology. They did. Upper Palaeolithic kilns are known from the Czech Republic, for instance, but they were used not for pots but for firing figurines and clay pellets.

Well, of course they didn't have pots. Pots are heavy. If

one of the constraints on the size of your family is your ability to lug small children across the permafrost, you're hardly likely to load a dinner service. I think back guiltily to all the stuff I have back at home: thousands of books, hundreds of notebooks, hundreds of artefacts to which some memory or allusion has been uploaded.

The importance of all this stuff troubles me. Not because I'm politically queasy about so much possession in a world of poverty (though I am), but because it calls into question what I'd be without it. Perhaps so much of me has been outsourced to databases or memory banks of various sorts that, if it all went up in flames, I would too.

This goes crucially to the inquiry into the Upper Palaeolithic. How can I think that I'm walking a moor in an Upper Palaeolithic way if most of me is on bookshelves and hanging on walls in Oxford? Indeed, doesn't it go further than that? Doesn't it go to the basic question of consciousness, of subjectivity, with which I've been so obsessed so far in this book? For much of what I say *I'm* saying actually resides a long way outside my head. I assert things with the authority of books that I know I've read, which I'd have to stand on a stool to reach, but whose actual contents I cannot consciously remember. My asserted authority is bogus. It's not even the fraudulent authority of the plagiarist: at least he knows and has consciously adopted what's he's stealing.

What must it be like to walk from this wood to the sea with nothing except what is really in one's own head, and really under one's own feet, and really in one's lines of sight, smell, hearing and so on? That's authority: that's dependency, and so that's elegance and self-possession.

I can't imagine it.

But I can try. One way is to be more nomadic. I've told myself that Upper Palaeolithic people weren't constantly on the move: that they'd camp for a while if there were no compelling reasons to move on, if there were caribou and hazelnuts enough where they were. That's true, but I've been using it as an excuse for being Neolithic. The truth is that I've become attached to my little beech tree bower. I like the mottling of my bag in the early morning sun, the gag in the cuckoo's throat just before the '*cuckoo*' comes, the flat stone I use for frying strips of pigeon, the angle of the shelter's main strut (which I notice I like because it looks *permanent*), the wheeze of the distant sea, the welcoming parting of the nettles at the edge of the wood where the path I've beaten begins, the moss that looks like a three-headed gecko, the wheeling buzzards nesting in the next valley and the reassuring weight of my rucksack.

All these attachments undercut the whole enterprise, and in a fit of iconoclastic zeal I spring up, tear down the shelter, throw the flat stone into the blackberry bushes and everything else into my rucksack, lace my boots as if I'm going to walk in them rather than sleep in them, sling the rucksack on my back, decide it's too heavy, take it off again, dump an old army blanket I've been carrying 'just in case' and head off in a direction I've never taken before.

It had to be done like that, explosively. If it hadn't been done that way, it would never have been done at all.

★

I'm on my own, and walking, walking, walking, diving down into the dark, cool pubic valleys whenever I can, distrusting the big vistas of the high moor, for from up high I can't help seeing vehicles threading along the roads, ships sliding, planes violating the blue. I've taken off my boots and put them away at the bottom of the pack so that I'll be less tempted to put them on again. In fact it's not a temptation at all: I wonder why I usually deprive myself of the knowledge that my feet are now giving me – knowledge that comes (like most knowledge worth having) with a *frisson* of sensation that never abates.

The air in the valleys is thick with red deer musk and spiralling clouds of tiny flies. The canopies are turning sunlight to sugar and pumping it through thin stiff veins. I'd prefer to sleep here, but the midges would suck me dry, and so I'm always on the moor at night.

I sleep when my legs get tired, and so the days get longer as I get fitter. I eat as I walk. There's plenty at the moment, for whortleberries and blackberries have come early, and there's plenty of wild salad. I assume that walkers are out to hunt and kill me.

The map is in my head from previous trips, I'm afraid. I should have gone to Siberia or anywhere else I didn't know well. For maps – as opposed to real knowledge of the land – are to do with dominion. They're literally reductionist: they reduce miles to centimetres, chatty and ineffable woods to splodges of green. And they give us the idea that we can fold up a landscape and stick it in our pocket. The land becomes about us. Maps are the worst example of what symbolising can do.

But despite the mental map, the moor triumphs over my idea of it. That stream there isn't on my mental map; nor is the barbed twig that rammed up between two of my toes and had to be dragged out with a big lump of me on it. And at night, or whenever my head is below the level of the grass or the heather, the map dissolves. The moor from down here – the moor of the fox and the mouse – has a completely different set of co-ordinates from the moor as seen by sky gods, mapmakers and reductionists.

I'm spiralling out from the shelter in rough circles of constantly increasing radius, eventually walking twenty or so slow, swinging miles a day. The sun shines relentlessly, trying to be kind. In the night sky there are only pursuits and dances: no constellation strides alone through deep space. X usually camps a few hundred yards from me. I can smell the smoke from his fire, and sometimes roasting meat. He's a proper hunter.

Sometimes I have to cross a road, and I do so fearfully, watching it for a while for fear of meeting a car. I'm afraid of the engine's roar – so much louder than anything in the rest of my life other than thunder – but even more of getting a blast from the car radio, or a snatch of talk. Up here there are no lies, and everything matters profoundly and for ever. Everything's a matter of life and death – as it is in the cars too, of course – but up here everything knows it.

I'd be a scary sight to the people in the car. I'm hairy, matted and literally wild-eyed. I imagine people looking like me roamed Devon after the First World War, unable to sleep for what they'd seen and lost, wondering how any one world could accommodate both Passchendaele and a nightjar. That's the root of my disorientation too. For in a few weeks alone on the moor, I have, despite all my misgivings, got somewhere with this Upper Palaeolithic project, and an echoing chasm has opened between the world of the moor and the world of the road – the road to which I know I'll have to return.

The emperor Akbar the Great caused this to be written on the gate to the mosque next to his palace in Fatehpur Sikri, in Uttar Pradesh: 'Jesus, Son of Mary (on whom be peace), said: The World is a Bridge, pass over it, but build no houses upon it.'

'The whole world is nothing but a very narrow bridge,' said Rebbe Nachman of Breslov, 'and the main thing is not to be afraid.'

It is houses that make us afraid, I've learned.

That's a problem, because my family live in one.

★

It's beginning to feel like autumn. Not always. There's nothing to see: no browning or falling leaves. It's not that it's cold. Rather, there's an absence of heat, as if the flames have gone and the world is cooking on old embers.

And then, suddenly, and on the high top of the moor first, autumn comes. But it is not a coming. It is a retreat. Autumn and winter aren't anything themselves. They are an absence of summer.

'Now that', said my friend Kate, a psychiatrist, 'is as clear a sign of a depressive personality as you'll get.'

Not so, Kate. It's not depressive to get your thermodynamics right.

I hate the fraudulent blandishments of autumn. How it tries to pass off the work of spring and summer as its own. Those apples, plums and berries, let's be clear, are nothing whatever to do with you, autumn: right? Your sole contribution to the harvest is to make the weather dubious and the tracks muddy. 'Mellow fruitfulness'? Yes, but not yours.

The family has gone back to what they call home. Meekly and unreflectively we send our children to schools, and they're about to start.

I'm still up on the moor, but now it sucks the heat parasitically out of me when I lie or sit. The mist that crawls out of the valleys at night doesn't crawl back with the day. I have torrential diarrhoea from all those wild plums, and it's getting painfully cold to squat in the streams to clean myself.

I don't *decide* to go: it just happens, like most things. I spend a night watching foxes skirt round pools of moonlight, listening to the drawl of the sea and smelling wet heather and coal tar on the wind, and then, somehow, I'm in a bus lurching towards Taunton, and there's rain streaming down the windows, and someone on the radio is lamenting, using two electric chords, the behaviour of some girl he's just met, as if it matters.

AUTUMN

'[I]n Sanskrit, which is the great spiritual language of
the world, there are three terms that represent the brink,
the jumping-off place to the ocean of transcendence:
Sat-Chit-Ananda. The word "Sat" means being. "Chit"
means consciousness. "Ananda" means bliss or rapture. I
thought, "I don't know whether my consciousness is proper
consciousness or not; I don't know whether what I know of
my being is my proper being or not; but I do know where
my rapture is. So let me hang on to rapture, and that
will bring me both my consciousness and my being."'
Joseph Campbell, cited in *Joseph Campbell and the Power of Myth*

'Achilles can walk in Altjira. Indeed he must: he has such
a lot to remember. Not least of the memories is that to
live as a human being is in itself a religious act ...'
Alan Garner, 'Achilles in Altjira'

Once you start walking, it's hard to stop.

I can't settle, and after a few pathetic weeks of trying to
feel like a nomadic hunter–gatherer by going for long walks
from our suburban house, fasting, breaking the fast by eating
fruit and having achingly cold barbecues in the back garden,
I get the boat to Bilbao.

There is some method in this self-indulgent madness.
It is that I'm searching for the origins of symbolism, the
most potent symbols are words and the Basque language is
thought to be one of the few surviving Proto-Indo-European

languages ('PIE'). Basque's ancestor probably originated in Anatolia and spread from there, evolving as it went. Basque may be, then, a living Upper Palaeolithic fossil with Turkish blood. To hear cavemen talking you might just have to go to Spain and sit in a bar with your eyes shut and your imagination racing.

So that's what I'm doing. I've been in an old dark place near the docks for the last four hours or so. It smells of sweat and seaweed. My table was varnished before the war and wiped just afterwards. I'm in the far corner, in a little crate made of Arabesque tiles and fish boxes. The mournful waiter fills up my jug of red wine without me asking, feeds me with octopus and calf's foot jelly, and tugs on his moustaches, bowing as he backs away.

The place was empty when I arrived. Now it's packed with off-duty pimps and stevedores. Very little light makes its way through the cigar smoke into my corner. Faces sometimes loom through the smoke towards me, see that I'm a foreigner, grunt and withdraw.

Everyone is speaking Basque. It would be dangerous to speak anything else. It's like nothing else I've ever heard: guttural, sibilant, hissing and rolling, like Arabic would sound if spoken by a Tunisian legionary whose first language was Latin. It feels extravagantly hyper-expressive, the language of a people drunk on words and their power. I can believe that that's how you'd feel about words if you'd met them for the first time. Language is truly magical: it can transfer things invisibly from one head to another with breathtaking precision. It can conjure something up (an elephant, for instance), with a single utterance. If you don't get excited about that, there's something wrong with you. And it's not surprising that the very earliest languages would want to push the boundaries; would want to celebrate their own power in every cadence.

The arguments for Basque being a remnant of PIE are complex but convincing. Some of the most immediately attractive arguments are controversial – such as the suggestion that the Basque words for 'axe' (*aizkora*), 'hoe' (*aitzur*) and 'knife' (*aizto*) are rooted in the word for 'stone' (*haitz*) – and hence go back to a time when axes, hoes and knives were made of stone. I hope that's true, but the Basque/PIE thesis can survive happily without it.

The structure of the language is very odd indeed. The central nouns have unusual prominence: definite and indefinite articles, which often elbow aside the nouns to which they refer, are thrust into their proper place. It gives the impression of a delight in the concrete world – of entrancement with the extraordinary fact that a word can represent anything truly. Verbs are shifted, as in German, to the end of the sentence: action is dependent on actors. (I wonder if that's really what's going on when Germans capitalise all their nouns.) Actors are primary.

Thus, in Basque, 'The man falls in front of the bear' would be: 'Man-the bear-the in front fall is.' 'The woman has given the berries to the child' becomes: 'Woman-the child-the berries-the given has.' And 'The hunter has seen the wolf' is: 'Hunter-the wolf-the seen has.' It's a pattern that emphasises inalienable personal responsibility for action. There's no hiding shiftily behind a 'The'.

*

Back I came, chundering through the Bay of Biscay, thinking that I'd learned something, and thinking that I should go back to the Derbyshire wood to see if I had. But the rain came, and the rain stayed, and so, for a while did I.

I'm sitting in a medieval library in Oxford. On the table in front of me is a big illustrated book about prehistoric art. Two plates obsess me. I've looked at them for many hours.

The first is the lion man – an upright figure with the body and limbs of a man and the head of a lion – made from a piece of mammoth tusk and found in the Hohlenstein-Stadel cave in southern Germany. It's between 30,000 and 32,000 years old.

The lion, to my eye, has a faint, ironic smile, rather like the Archaic smiles – lip smiles only – on the faces of the Greek *kouroi*. He has a broad, big-lung-ed chest and powerful, spear-throwers' shoulders. Some say he's stiff, but it's the balletic stiffness of a paused stride. In this very early period of human art, human male bodies always had animal heads. Women were always fully human – though their breasts were often huge, their vulvas dramatic, their bellies and pelvises capacious and their thighs like oak trees.

As we've seen, the most usual interpretation of the animal–human hybrids is that they represent some aspect of the shamanic world – perhaps shamans in the process of transforming into spirit–animals or vice versa: that the borders between worlds and categories are open; that this world is not all that there is; that things are not as they seem.

The female figurines tend to be read as celebrations of the life force. They may be mother goddesses, or fertility icons, or both. They say that the universe *gives*, and accordingly that humans are not self-created, their stories are not self-authored and their disposition should be gratitude.

The other plate is of the lion cave from Chauvet, in the Ardèche region of France. It shows a pride of mane-less cave lions hunting a herd of bison, and it is at once the most beautiful, adept and frightening work of art I know. It too is between 30,000 and 32,000 years old.

The lions have no back legs: they would distract from the terrifying directionality of the forepaws and the heads, which all point towards the bison. It's as if some great cosmic finger is pointing at the bison. This is a painting about being singled out for death. The *focus* of the lions makes me shiver. The eye of the most completely rendered lion is invisible, but is hinted at darkly. We know what it would look like because we've seen it in nightmares. The bison are more suggested than drawn: they're all tangle and thrown heads. There's more movement on that cave wall than in a typhoon.

This was done by someone who had been inside the natural world and had no illusions about it. But he (it tends to be assumed that it was a he) was not just inside the natural world but outside it too. He, like electrons, had non-locality. And he could not just observe and record events; he could tell a story. Those lions came from somewhere, with histories, and things were going to be different because of the killing that was about to happen. This story, as shown by the hidden eye, was part of a bigger one – a story in which the painter had a part. This wasn't just about the death of a bison; it was about Death.

Yet it is terribly beautiful. The beauty is not in the terror, nor the poetry in the pity. The artist does not merely admire the lions as killing machines. He admires them because, regardless of their function, they are intrinsically beautiful. Nature, for this artist, is not just a larder.

Behavioural modernism, the archaeologists tend to say, is not only evidenced by symbolism; it's all about symbolism. The lion man and the lion cave are supreme and early examples of this symbolism. But as I've reflected on them and on my experiences in woods, seas, friendships, the African bush and on mountains, moorlands, beaches and shamans' couches, I've become heretical. I've come to wonder whether symbolism is all it's cracked up to be, and in particular whether its use really is the great watershed separating us from everything else that had gone before. The great divide, of course, is supposed to be between anatomically modern but behaviourally non-modern humans (whom we can see in the fossil record for about 150,000 years before *we* – announced by extravagant symbolising – made our grand entrance into history).

A few years ago I was in Namibia, following a big bull kudu. (I'd been trying to write about rogue males and romantically, on slender evidence, had got the idea that this was one of them.) Tjipaha, a Namibian tracker, was with me. He was a grizzled man of about forty-five. We'd been on the trail of this animal since dawn. It was now late afternoon. Night would come quickly, like a dropped curtain, and we didn't have much time left.

It was the dry season, and there were very few tracks that I could see. We'd had to rely (so I gathered) on broken stems of grass and faintly disturbed leaves. But we came to a damp patch, and there, in the mud, were tracks even I couldn't miss. Not just of our kudu, but of other kudu too, and warthog, blesbok, hartebeest, wildebeest, dik-dik and impala, all embedded in a lacework of small mammal and bird prints. Tjipaha looked hard at them for several minutes, walking round them all, crouching to have a closer look, sniffing, then swivelling around to look in the surrounding acacia scrub. He walked away, came back and crouched again. Then he stood up, brushed himself down and began to speak slowly but unhesitatingly, and with complete confidence.

This kudu, he said, had been at this water three nights ago. That was its first trip here for a while, for it had been travelling from the hills in the east. That night it had stumbled slightly as it rushed up the bank after being spooked by a frightened hartebeest. It had come down on its right 'knee' and had found it difficult to get up. It had spent an uneasy night under a thorn tree, unable to settle, fearful of moving too far on the aching leg. In the morning it had returned to the water, chasing off some sand grouse, and had then browsed for most of the day in a grove a mile away. Coming back, still jumpy, it had been stampeded by a lame jackal, and had turned back the way it had come – towards us, but going first for a drink at the rivulet – and the next day had retraced its steps again to the grove. Another drink, another night, supping water beside a cobra; another night under the thorn tree before turning back to us yet again, its leg better, but still carrying slightly too much on the left forelimb – which wasn't easy because it had been kicked in the right hind limb by another kudu a month or so before. That was when we picked up its spoor.

I didn't believe a word of it. I thought it was play-acting. I'm not sure I believe it now. But he talked me through it all, offended by my scepticism, and if it was all a performance, it was a compelling, detailed and perfectly reasoned one.

Even if he was having me on, there are certainly people who can do this, and there were very many more people who could do it before supermarkets and rifles atrophied the necessary senses and intuitions. But also before behavioural modernity. Pre-behaviourally modern humans were sophisticated and extremely effective hunters. They must have been able to read the ground at least as well as my tracker.

Now here's the point: that sort of reading demands an appreciation of symbolism of precisely the same kind, if not to the same degree, as the symbolism we see in the lion man and the lion cave. Good trackers have to know that something stands for something other than itself. This is obviously true on a very basic level: a footprint is not the same as the foot that made it. It is a symbol of that foot. Good trackers take this symbolic mode of thinking to a supreme level, knowing that if there's pollen in a footprint, the footprint must have been made before Tuesday afternoon because it was only on Tuesday afternoon that the wind was blowing hard enough in a north-westerly direction to spill the pollen. Pollen – which is really the ejaculate of flowers – becomes the symbol of a window of time. The clairvoyant ability to track goes well beyond mere appreciation of the architecture of webs of causation. It is not materially different from understanding that a smear of charcoal paint on a piece of stone can become a lion's throat. We see evidence for symbolisation more or less whenever we see evidence of the successful hunting of animals that would have needed to be stalked over long periods. Whenever we see hunting expeditions, we see symbolising. Every prehistoric barbecue was behaviourally *modern*.

★

At last the rain eased off, and on a grim late September evening I watched myself walking up the dale, past the cottages built for the lead miners, past the superb pub (very painful to walk by), past a chapel built on the assumption of a fulminating God and now deserted because God has calmed down or died, up into the lead field with its galloping steers.

The wood is mourning-band black, with a thorny back. As I push open the iron gate that leads in, the wood stops breathing and starts watching. It has frozen, with one forepaw held in the air.

Tom wouldn't come. 'I've got a project I need to do. And a guitar lesson.'

'You can do the project when we get back. We needn't be there long. And take your guitar and practise all day and night.'

'No: I need to be here.'

Very wise. Though I told myself that it was nothing personal, I've really no idea what was going on the last night we were here, and nothing that's happened since makes me suppose the wood will be safe now.

I climb up the hill, as if up a scaffold. I won't go back to our original shelter. Perhaps what happened was very local to that place. It might be better if I'm not under the trees too. Out in the open I'll be able to see comforting planes and satellites.

Sarah's house is empty. She's in London.

I lay out my waterproof bivvy bag on a flattish area of grass, put a sleeping bag inside it, get inside, eat a can of sardines and lie back.

There's nothing much to hear. The wood still hasn't breathed. Sometimes there's a distant car with a broken silencer. Nor is there much to see: just the hunched trees down the hill and the lights of a farm a couple of miles off. No Upper Palaeolithic hunters in caribou-skin cloaks by the barn. And there's no coal tar soap at all.

I wait. The wood waits. It seems as if we're playing chicken. Who will breathe or blink first? What will happen when one of us does?

I never find out. The wood can wait, and has the will for it. I don't have the nerve to breathe, blink and say: 'Right: what now?' I've wondered ever since what would have happened if I had.

I stare into the night until the sun comes up. And now I discover something very unexpected. I'd assumed that the wood was testing me. And so it was. I'd presumed that I'd failed, and that I'd have to creep back down the dale with my tail between my legs and get the train to Oxford – probably in time for Tom's guitar lesson. But it seems that it wasn't a pass/fail examination. I might not have got the first-class honours of epiphanic union with the hill, but I'm not sent home empty-handed either.

The hill is as generous as Sinai and my friend Chris had

been all those years ago. It doesn't take umbrage, saying that I'd been off committing adultery with Exmoor and so wasn't welcome back to the family home. It sends along the tick-tocking magpie, and there is, again, a one-eyed robin – though I'm not sure it's ours: this one looks more subdued, but perhaps it's just been beaten up by life. And then, *by day!* – as a supreme sign of grace and forgiveness – the hare. I'm sitting in the corner of the field and she comes almost to my feet and she looks at me and I look at her and the circuit between us is complete and something goes on that will never go off.

So I stay, sleeping by day and travelling by night because night is when the land comes out and can be met. I walk long distances, often staying away from the wood, sleeping in ditches, in the shadow of drystone walls, in caves on gritstone edges, under piles of dry leaves, on pine needle mattresses and on the moor with the lights of Sheffield in one eye and the glow of Manchester in the other. With cold fat grouse and owls so soft and foxes so hard that the rising cold bounces away off them.

The last swallows are gone, and the geese are starting to come back. In the last few weeks of the summer I buzzed frantically around, trying to soak up the last of the sun, gathering the last of the nectar, knowing that it would have to last a while. But now I'm not worried about the cold. It's not because there's plenty of fat around my soul. I feel what before I've merely known: that the land endures and will come again, and if I can stay as close to it as I am now, I will endure too.

I'm circling now; circling closer and closer to Sheffield, sleeping now on old slag heaps and in scraps of coppice where the city gives up and the wild begins. Sometimes I'm close enough to the roads to see buses go past, and I wonder if there's anyone on the bus that I was at school with, and if so how their life and mine would be different if we'd kept in touch. This should disinter the old allegations of betrayal: 'You left them. You left here. You're a fake. Just think what you might have been if you hadn't cut the roots, going deep into this place, that fed you. You might have been a proper tree by now.' But they don't come.

X is here all the time now. Often very near. Sometimes his face is half-turned to mine. Sometimes I catch his smell – sweat, urea and wood smoke – when the wind turns. Usually, though, he is just a grey bulk at the edge of my mind.

For days now I've assumed that I'll end up at Sinai. I can sometimes see it, above the trees above the suburbs above the river of brown moor water. But one evening, waking up in the cleft of a tree where I've spent the day, I know that I don't have to go there for forgiveness, or completion, or anything at all.

This is still a place haunted by regret and remorse and the smell of coal tar soap. But as long as I sleep out in my bivvy bag, the ghosts don't taunt or split me open with longing.

★

'Tell me a story, Dad,' says Tom.

We're back in Oxford, sitting on tree stumps in our communal wood near the nursery school. We've lit a fire, and we're frying sausages and mushrooms and hoping that the local jobsworth won't come whining about health and safety.

Very well, Tom, I'll try.

Once upon a time, when the world was already very old, a man and a woman lived here – there, by that oak tree. They had some children, and they loved them very much. Loving someone very much is always a problem, but this man and woman had a particular problem, because these children, like all children, had to be fed, and that meant killing other things that the man and the woman loved: animals and plants. Whenever the man or the woman lifted up their hands to kill an animal, or tugged on a stem or a berry, they'd hear a pleading; a screaming: 'Don't kill me. Please don't kill me. If you do my children will get you, and we'll haunt you for ever: just see if we don't.'

What were the man and the woman to do? They couldn't let their own children starve, but they couldn't bring themselves to do the killing.

Their own children grew thinner and thinner. Their ribs showed, and their cheek bones stuck out of their faces.

Then one day an old woman limped into the wood.

'What's wrong with your leg?' said the man. The old woman lifted up her long skin cloak, and the man saw that a long thorn was sticking into the leg. The wound was infected, and flies were buzzing all around it.

'Sit down,' said the man, and he and his wife pulled out the thorn, bathed the wound with water from the stream and bound lichen around it.

'You're very kind,' said the old woman. 'Thank you. And now I'm very hungry. Do you have something I can eat?'

The man and his wife looked at one another. They were very embarrassed that they had nothing to give to the old woman.

'I'm very sorry,' said the wife, 'we have nothing. That's why we're all so thin.' And she told the story.

'Dear, dear,' said the old woman. 'That won't do. Let me see if I can help.' She shut her eyes, counted slowly to three and, on the count of three, whooshed though the roof of the hut.

The man and his wife were astonished. Where could she be? They didn't have long to wait. In a couple of minutes the old woman appeared again, sitting on the ground in the middle of the hut.

'It's going to be all right,' she said. 'I've spoken to the animal gods and the plant gods, and they're happy that you eat their animals and plants as long as you're kind.'

'That's a relief,' said the man. 'Thank you. But where did you go?'

'Just to the place where the plants and the animals come from, and where they'll go eventually, and where their gods live.'

'Couldn't we go there too?' asked the woman. 'We'd like to say thank you, wouldn't we?' And she looked at the man, who nodded.

'Well, you could,' said the old woman, doubtfully. 'And one day you will. But are you sure you want to go now?'

The man and the woman were frightened, but said that they did, and the old woman led them out through the wood to a place where there was an opening in the rock. The man and the woman had always been frightened of this place, because they knew that bears lived there, but the old woman said something to her stick, and it lit up, and she led them in and down, and further down.

I don't know what the man and the woman saw down there. They would never tell me. But what I do know is that from that day their children were fat and strong, and that the man and the woman, and the children too when they were older, went often to the opening in the rock when it was night, and when they came back they looked different.

The family became famous around here. There were two reasons. The first was that, as soon as the old woman had left, the man and the woman went to the place in the wood where they had buried their parents, years before. They scratched and they scraped until they got to the bones, and then they washed the bones carefully and put them in their hut. And when they went on long hunting trips, as they sometimes did, they would each put on a bracelet made of hand bones, leaving the rest of the bones safely under a rock until they came back.

The second reason was that they became great storytellers. In the winter they built huge fires. They had to be huge because everyone from round about would come and sit by the fire to listen to the stories. There had been a kind of story before, of course, but they were more like reports of what had happened: how the deer hunt had gone; which berries had made them sick. The stories that the man and the woman told were different. They were travel stories, but of travel to strange places and meetings with strange beings, and of the gifts that they brought back, and about how if you looked hard enough at a flower, or at anything, you'd see that things weren't always as they seemed. And they told them as if they mattered, and as if there was a point in everyone finding and telling their own stories, because those stories, if not the people themselves, would go on for ever. Sometimes the stories became so wonderful that words weren't enough, and then they'd burst into song, often in the middle of an inadequate sentence or an inadequate word, or get up and dance around the fire, showing how those beings had walked or strutted or slunk.

The man and the woman got old, but they didn't seem to mind. The time came for them to die, and they told the children, who were now the biggest and the fattest in the clan, that after the birds had cleaned them off, their bones should be put on a high ledge in the cave the old woman had showed them all those years ago, together with the bones of their own mothers and fathers. And this was done.

'Very interesting,' said Tom. 'And now I really do have to play my guitar.'

★

'You're back from mooching with the archetypes, I see,' said the highly scented woman.

Am I? I'm not sure I *am*, to be honest.

NEOLITHIC

WINTER

*'I am not decrying the profession of accountancy, only its
appropriation of competence in every field. And if, as it
looms, we are entering on a period biased toward materialism
at the expense of progress, then we are in the hands of the
accountant, a spiritual Ice Age, where all will be frozen
and there will be no risk, and without risk, no movement,
and without movement, no seeking, and without seeking,
no future. Darkness will be upon the face of the deep.'*
Alan Garner, 'Aback of Beyond'

When you stop walking, things happen.

I'm in northern Kenya, in the highlands not far from
Mount Kenya. I'm staying with a zoologist friend who has a
farm there. It's midday. I've been up since well before dawn.
I was woken by a banging on my bedroom door.

'Please, please, up. Fire. Fire.'

As soon as I swung out of bed I smelt it: smoke too sweet
to be frightening.

There was no reason to panic. It was a bush fire, several
miles away, but a gentle breeze was blowing towards us. We
sat on the terrace, ate pawpaw, watched the giraffe at the
waterhole below the house and from time to time picked up
the binoculars and looked at the fire. It was just a thin line of
blue cloud, not threatening, not visibly advancing. But the
giraffe sometimes levered up its head from the water, swung
it round and stared at the line. I went back to the pawpaw
and my book.

But now we've driven towards the fire and parked the Land Rover on a hill half a mile away from the front line. The bush below us is heaving with wildlife: eland, duiker, Thomson's gazelle, warthog, buffalo, oryx and birds that prefer to walk. They're all moving towards the house, stopping often to look back over their shoulders. We drive towards the fire. We hear crackling now. Just behind the smoke martins swoop and dart, picking off the insects driven up by the flames.

The flames themselves aren't impressive. There are only occasional flurries of intense blaze when the fire reaches a whistling thorn. Then the bush does whistle as hot air eddies in and out of the black gall sounding-chambers, and there's the odd squawk as a branch is ripped from the trunk. But mostly it's a fag-ash creep through the brown grass.

We're not worried. A fire like this is unlikely to jump the dirt road that cuts between this area of bush and the farm, and anyway the wind is starting to turn. We drive back to the house for the third pot of coffee.

Overnight the wind blew the fire back into itself. It had already eaten everything, and soon starved.

Next morning we walk into the black wasteland. It's still very warm underfoot. The jackals, hyenas, vultures and crows haven't left much behind, but there's still a Sunday lunch smell of roast meat. Sometimes the remaining bodies are recognisable. There are lots of four-toed hedgehogs, some of them still curled up, thinking their spines would see off the flames, left by the scavengers because, for the moment, there are easier pickings. A brush-tailed porcupine that just couldn't shamble fast enough lies on its back with its front legs begging and its lips pulled back in a last snarl. But mostly the corpses are anonymous – charred tubes that in life would have been spiny mice, grass rats, and ground squirrels. There are many snakes, some of them knotted or curved unanatomically tight, some straight and stiff like ebony broom handles. The very small mammals may have survived, squeezed into holes between the grass roots which are already preparing to push up into the light.

Back at the house I wondered how I would have felt if I'd started that fire.

★

The Neolithic was the age of domestication – of cereals, sheep, goats, cows, pigs and us. But before all that came the domestication of fire. *Homo erectus* and *Homo neanderthalensis* had tamed fire. It had been flung down from the heavens as lightning, or oozed out of the water as flaming marsh gas, or sparked out of struck stone and caged in moss or fungus, or lay silent and invisible in the flint, ready to be summoned by a blow.

It had done great service. It had seen off lions, unlocked otherwise inaccessible calories from food and enabled the building of bigger brains and bodies, out-competed the sun and extended the day, and, as hunter–gatherers clustered around it, become the frame on which community hung. It may have spawned fecund metaphors which, once alive, would never stop breeding. These were all portents of a more hubristic type of mastery. For fire came to be used as an indiscriminate weapon against the natural world – a weapon of mass destruction. That use began to drive a wedge between humans and the rest of the natural world.

Prehistoric humans certainly saw what fire could do on a scale bigger than the hearth. They watched bush fires, as I did, and saw that, after the fire, game came to graze in the clearings on the newly springing grass. No doubt they thought that fire could bring meat closer to home, and no doubt fire was sometimes used to do that before the Neolithic. But by and large, hunting in the Upper Palaeolithic and the Mesolithic was an uneasy, religious business, hedged round with taboo. Taking a life was serious and risky. To avoid dreadful consequences to the killer and the community, killing had to be thoughtful, prayerful, targeted and liturgical. Yes, there were mass killings. Upper Palaeolithic hunters cut swathes through the big animals, contributing significantly (along with climate change) to the extinction of the characteristic megafauna of the Pleistocene. But the over-killing is more likely to have been due to ecological mis-calculation than to the sort of psychopathic detachment that came later.

In the Neolithic, fire was used to destroy everything that got in the way of man's convenience. In the olden days to break a tree branch demanded a request and probably some sort of strenuous contrition. Now whole trees, whole tracts of land, along with all the communities that lived there, went up in smoke. The old laws had been broken, and on such a scale that the old means of reparation and propitiation were unworkable. Someone who'd destroyed a forest couldn't go on a shamanic journey to heal his scorched soul: he'd be lynched by all the outraged spirit animals.

So, yes, I'm suggesting that the simple use of fire to clear ground necessitated a massive theological shift. The individual souls of the destroyed organisms could not possibly be appeased. It became necessary to switch allegiance to a transcendental being, overlord of the devastated swathes, who alone could authorise the destruction. This switch took time. Ancestor cults were a step on the way.

Once you start killing indiscriminately, you become an indiscriminate killer. Acts change identity. Thus humans, by the act of putting a torch to brushwood, reinvented themselves. They had been ontological equals of the deer and the trees: now they (perhaps under, eventually, a feudal God who delegated authority) were their masters. Fire and felling took out much of the luxuriant woodland that covered Mesolithic Europe. But their effect on humans was far greater. Acts of violence usually affect the inflicter more than the victim. It may take some time, but the universe always gets its own back.

We only want to burn the land if we're fed up of wandering to get food and want to eat from the wood or the field or the shop just on our doorstep, or through the efforts of others. That desire for convenience is deadly. It puts us in breach of the immutable law that says that things start going wrong – for us and for everything we touch – as soon as we stop walking.

★

It's another early morning. The alarm clock goes off at 3.30. I splash water in my face, dress, jump on my bike and cycle off. It is bitingly, cruelly cold. I see no one at all until I get to the minibus. Most of the others are in already. Like me, they don't want to go to the slaughterhouse. I nod grimly at them. The ones who are awake nod back.

'Fun and games today.'

The journey takes about an hour, over flat, frozen turnip fields. We get to the slaughterhouse before five. It stands by itself, a couple of miles from the nearest town. It's already busy there. It hums and clanks and even outside, with frozen noses, it smells of pig shit. Several lorry loads of pigs have arrived. From the addresses on the lorries, those pigs had an even earlier morning than me. Some of them have come from pig plants 100 miles to the north – great plastic hangars with no natural light. Their feet have never touched the earth; their wet, quizzical noses have never snuffled anything more interesting than the biochemically optimised concentrate that a robot squirts into their troughs.

The manager comes out to greet us. He's in a suit and tie and short white clinical boots. A polythene bonnet stops some of the dandruff falling onto his shoulders. He shakes our hands damply. He has pudgy fingers and immaculate nails.

'Welcome. Welcome all. We're very glad that you've come to see how it's done properly. Any veggies here?' He looks around, smiling tolerantly.

'No? Good. Not that it would be a problem if there were. Nothing to hide here. Nothing at all. I'm going to hand you into the tender care of Ron. Ron's our senior foreman, and every one of you here will have had a pie filled by Ron's handiwork. Isn't that right, Ron?'

'Must be, sir, by now.'

Ron is in his late fifties, white martial moustache, dragon coiled round his enormous forearms, other self-drawn tattoos on his knuckles, nothing behind the eyes; he's worked in an abattoir, with breaks, since he left the school down the road.

'You can leave your kit in here,' and he opens the door to the 'Workers' Lounge'.

'*Kit*', I note: he's trying to sound military, but I'm fairly certain he's never taken a shilling from Her Majesty.

The lounge smells of stale breath and farts. The bin is full of Coke cans and chips. On the table in the middle of the room there's a big stack of porn. One magazine is open at 'Readers' Wives'. 'Sorry about that,' says the solicitous Ron, closing it up and putting it back on the pile. 'The lads need to unwind. It can be stressful, this job.'

Around the room are the lockers where the workers leave their things. They've got their names on: white names: Barry, Gary, Len, Steve. 'Steve' has been crossed out and replaced, in big capitals, by 'PEEDO'.

We dump our bags, change silently into the boiler suits and boots Ron throws us, and file through into the business part of the slaughterhouse. It is cold and quiet. There are metallic clinks, the whoosh of hoses, the swish of sharpening knives and, from the far end, a stirring under some wisps of steam.

The stirring is the pigs. They're in holding pens now, clustered together for warmth, waiting an hour or so to calm down after the journey. They're not distressed or disoriented. They have never had any pleasures; they have no expectations to dash, and so far there is no blood to smell or screaming to hear. When everything's ready at the sharp end, they will be nudged up a high-walled passage by Steve the Peedo holding a board.

We're allowed into the holding pens, just to show us how happy everything is, and we scratch some backs and the pigs look up at us with those little eyes that, per square centimetre of exposed eye surface, are the most expressive in the world – and that includes the Royal Shakespeare Company's finest. No human has ever scratched their backs before, and no human will again.

They're calm enough, the sharp end is ready and waiting, and Steve starts chivvying them. The pigs at the back get reluctantly to their feet, and start exploring their way up the channel as all pigs do, trying to stick their snouts under the roots and rocks that aren't there under the non-slip concrete floor. The pigs in the middle move because the pigs at the front are moving, and the frontline pigs are their friends, whose company is comforting in this strange place.

At the front, the first pigs are approaching the stunner. The first one arrives. The stunning crate isn't sinister to pigs used to galvanised bars. When the pig goes in, blunt jaws close behind it to stop it wriggling back. Barry reaches in, puts tongs over the pig's head, and presses a button. Current surges through the pig's head, its body stiffens, its eyes close as if it's being told a bedtime story, a trapdoor opens and the pig goes down a children's slide for the next phase.

This is more interesting than horrific to the pigs behind. The manager's right so far. I'd expected the reflexive screams you get when a pig is gripped by the sides, or a bit of death terror and stampede. It doesn't happen. The pigs and the men are on their best behaviour.

Happy with the performance so far, and wanting to quit while he's ahead, Ron moves us on with the stunned pigs.

I'm chastened by the lack of protest and ceremony. These pigs are not raging against the dying of the light. The abattoir is a business, and the pigs, outrageously, are behaving in a businesslike way. I blame them for their complicity. Death should be an event: this isn't. If death can happen this routinely, how can any of us ever feel safe? And as for ceremony? Well, these men got up, had their breakfast, had a shit, drove to the slaughterhouse listening to fake news and canned music on the radio, got changed, flicked languidly through 'Readers' Wives', hosed, honed, bantered and then stood there chewing, waiting for the 700 they'll kill today, just as they did yesterday and will do tomorrow and for ever until the end of the age, or until they get their heart attacks from too many bacon rolls, whichever is the earlier. What am I expecting from the men? The Lord's Prayer before each pig is stuck? CBT for the pigs near the stunner? Bereavement counselling for the pigs at the back at the queue?

The pigs are lying in a twitching pile at the bottom of the chute. Gary puts a chain round one back leg, and each pig is hoisted up, suspended from a slow-moving rail and moved to Len, the killer. He sticks a knife into the throat. When it's pulled out, it unstops a torrent of blood which gushes over Len's apron and boots.

I needn't have worried about the absence of ceremony. Len has a fine baritone voice, and he sings most splendidly, in time to the punches of the knife into the throats, 'All Things Bright and Beautiful'.

All [punch] things bright and beautiful,
All **creatures** [punch] great and small,
All [punch] things wise and wonderful,
The **Lord** [punch] God made them all.

Ron is proud of Len's wit, and grins and claps along, looking at us for approval.

The pig goes next into the scalding tank. Things can go wrong here. If Len's not done a proper job, the pig can regain consciousness, and when that happens, Dante himself would be lost for words: screeching, plunging, eyes rolling, swallowing gallons of boiling water, blood and dung. But today, at least, Len's on target, and the pigs process silently, twitchingly on to de-bristling, disembowelment and dismemberment.

Ron can breathe again. He delivers us back to the manager, job done.

'So, ladies and gents, nothing to be frightened of, is there? You can have your sausage and mash knowing that the pig's suffered no more than the potato, right?'

He's used that line before, and is rather proud of it.

No one says a word in the minibus on the way back. We're back by lunch. They'll start killing again at 1.30.

I hadn't met X at the time, but if he'd been there, I'd have done anything to avoid catching his eye.

The slaughterhouse was a very long way from the Upper Palaeolithic, but it's amazing how far you can get if you stop walking.

★

It is crass and unfair to judge a whole process by its end, particularly when the process lasts many thousands of years and takes place across the world, in many different cultural settings. We should be slow to link the (slow) Neolithic revolution with the evils of abattoirs, states, fast food, hedge funds, social alienation, the internal combustion engine, the subjugation of women, the class system, boardroom sycophancy and the annihilation of Amazon parrots. So I'll be slow. There: I've taken several deep breaths and can move on. Having been slow, and considered carefully the draft indictment and the evidence supporting it, I make the allegations aforesaid.

The causation looks like this. Humans (no, let's be honest, *we*) wanted convenience and what we saw as security. We wanted to reduce or eliminate contingency. We sought to rule the natural world, and began to see ourselves as distinct from it, rather than a part of it. Our early efforts at control were, in one sense only, very successful. We managed to produce lots of calories in one place. That caused a population explosion. Once population started to increase, there was no way back. We had to produce more calories, and to increase the size of the places in which we produced them. There was no escape from the places, or from the stern dialectic of Dunbar's number. Enter status, surplus, markets, big powerful men and hence smaller, less powerful men, and all sorts of camp followers, including overcrowding, loneliness, occupational disease, diseases of sedentary life and epidemics of infectious disease. Continue synergistically for 12,000 years or so, and you have us.

★

On the tundra where X and his son had hunted for ivory, danced with spirit wolves, watched the thoughts of their family hundreds of miles away in France and felt their Selves swelling inside their heads as a woman's belly swells with a child, the weather changed. The ice retreated north, the permafrost thawed and, fuelled by the dung of many millennia of mammoths and woolly rhinoceroses, forests burst out of the tundra. The natural world began to give even more lavishly.

★

It's an icy blue day, and we are at the Norfolk coast to look for grey seals, swim and drink parsnip wine in a friend's shed. We're walking along a bleak sandy strand. The kids are fighting, their alliances shifting like Beirut in the 1970s. The current dispute is about ownership of a cormorant's skull. I'm looking east, watching gannets diving. They're feeding on fish which are feeding on organisms who owe their lives to the sea bottom – to the submerged kingdom of Doggerland. This is the Mesolithic Atlantis, a lush paradise of oak, alder and hazel, trembling with edible animals, which was overwhelmed by the sea around 8,500 years ago. I went to give a talk at the Natural History Museum in Rotterdam, where many of the Doggerland artefacts, dragged up by trawlers off the coast of the Netherlands, are stored. To thank me for the talk they gave me, in a smart presentation box, a coprolite, squeezed out into Doggerland by a Mesolithic hyena. I *think* they were being kind.

Ahead of us on the beach there's a big black mass. The children, setting aside their differences for a moment, run up to it. It's a huge lump of very old peat, on the way to becoming coal: a piece of Doggerland, perhaps 10,000 years old. We can pull out individual twigs. A Mesolithic squirrel might well have scampered up that bit of alder.

The children pull out the branches and hold them as if they were the bones of saints. They stick their noses in until their faces are black, wanting to breath air that was last exhaled by a sabre-toothed tiger. And then, despite the cutting wind, they take off their coats to use as slings, loading them with as much peat as they can carry.

'I bet no one else in my class has a prehistoric forest,' says Rachel.

They do, of course, for there's lots of illicit coal burning round us, but I don't have the heart to tell her.

Back home they take out the Rotterdam coprolite and put it on top of the peat. They think this is very funny.

'It might have been *right there* all those years ago,' says Jamie. 'You never know.'

I suppose you don't.

★

The Mesolithic hunters who heard the wind in the branches that are now in our kitchen had brains that, as social facilitating devices and in other cognitive ways, worked just like Upper Palaeolithic brains, Neolithic brains, Bronze Age brains, Iron Age brains, Greek and Roman brains, Dark Age brains and medieval brains. We'll return to the question of whether they work just like ours.

Many would disagree with this, saying that settlement created the necessary and sufficient conditions for great strides in the use of symbolism, and thus for great changes in our cognitive architecture. This is a form of cognitive condescension, which sees Upper Palaeolithic humans as intellectually and spiritually rudimentary. The reason to reject it is not that it is politically incorrect, but that it's contrary to the evidence.

To understand that evidence we need to pick up the human story in the Levant around 14,000 years ago. There, in dense oak and terebinth woods at the edge of the Mediterranean, lived the Natufians, yet another example of a people who (along with most other peoples) defy the usual anthropological categories. Were they hunter–gatherers? Yes. An important time of their year was the arrival of the migrating herds of gazelle, just as the movement of caribou was crucial for the Upper Palaeolithic people of northern Europe. Wild ass and wild pig were important too, along with nuts, berries and roots. Did they wander all the time? They did not. Did they live in villages? They did. Did they have cereals? Indeed they did: they had been cutting wild cereals for thousands of years with flint sickles, but had never (or only rarely) domesticated them. Why should they? The plants self-seeded, giving enough for the following year without any of the downsides of cultivation. It was all rather Edenic. They ate bread, but not by the sweat of their brow.

Climate change expelled the Natufians from Eden: the fast geological cooling called the Younger Dryas, which, for well over a thousand years from around 12,900 years ago, locked up water in glaciers, creating a drought in the Near East, withering the oak, the terebinth trees and the grass (particularly the wild cereals) and depleting the herds of game and disrupting their previously clockwork movements.

The Natufians faced a serious challenge. What should they do? Their world, over (at most) a few decades, had changed beyond recognition.

They had two strategies: relocate and revert. They went, like their prey species, from the hills to the warmer valleys, where some of the trees had survived. There they built villages again, but they could not be as sedentary as they had been: they had to go back, at least to some extent, to the old wandering ways: to go to wherever the universe decided there was going to be food. For some – and particularly in Sinai and the Negev – this meant turning the clock back completely, and becoming committed hunter–gatherers again in the Mesolithic and Upper Palaeolithic modes.

Cultivation was difficult under these trying circumstances, but no doubt the exigencies prompted thoughts of raising cereals in the valleys, and perhaps there were some early attempts.

The Younger Dryas ended even faster than it had begun. Around 11,500 years ago global temperatures rose by 7°C in less than a decade, and the scene was set for the Neolithic. The old villages were repopulated and, from the shores of the Levant to the alluvial plains of southern Mesopotamia, new villages built. Here, at last, is the stable sedentism that we have all been taught was the catalyst for domestication, which (goes the propaganda) was in turn the catalyst for what we pejoratively call 'civilisation'.

This phase in human history has usually been represented like this: sedentism produces cultivated cereals (or perhaps cereals cultivated by some precocious agricultural entrepreneur in the Younger Dryas became widely available and facilitated sedentism). Sedentism and cereals increase the population. This increased population, under pressure from Robin Dunbar, needs to develop a particularly sophisticated social brain. It does, exploiting, multiplying and enhancing symbolism to do so.

There are two things wrong with this reconstruction. The first is the assumption that we *do* have better social brains as a result of sedentism. We'll look at that later. The second is the insistence that sophisticated symbolism and social life *require* a substrate of agricultural sedentism. This is contradicted not only by the superb efflorescence of Upper Palaeolithic art but also, supremely, by the northern Mesopotamian site of Göbekli Tepe.

You can reach it by getting buses for ever from the honking maelstrom of Istanbul bus station, and you should, really, rather than dropping suavely down to Sanliurfa's little airport. You should take lots of little dodgy buses, bald-tyred, full of smoke and goats, rather than the urbane air-conditioned ones with stuttering videos, and you should stop on the way at lots of little towns to eat kebabs and be stared at, for it's good to be reminded that the West is weird.

Sanliurfa, or Urfa, is an unlovely, optimistic town of Vodafone shops and spitted chickens, with islands of spicy darkness, and the desert at the door. Many years ago I tasted antiquity and politics for the first time near here, sleeping out on the banks of the young Euphrates, drinking sun-warmed red wine from my water bottle.

Göbekli Tepe is a short trundle in a minibus from Sanliurfa. It's an unlikely place to have all your anthropological preconceptions unpicked. There are no charismatic sphinxes winking conspiratorially with you as they let you into the joke they've played on academia. The highlight is a complex of T-shaped pillars, up to 5.5 m tall and weighing up to 16 tonnes, with human arms and hands on the sides – presumably showing that the pillars are human figures. Some wear belts and loincloths. Crawling, walking or flying all over the figurative bodies are non-human animals: foxes, lions, scorpions, boar, snakes, ducks and cranes. The site is vast, although little of it has so far been excavated. Huge numbers of bones (particularly gazelle, but also aurochs and wild ass) have been found. This was a place of great feasting. It may have had its own big brewery.

But it seems that nobody lived here. It was only occasionally occupied, as stable isotope analysis of the gazelle bones has shown. It appears to be a huge temple.

So far this is interesting. There are two things that make it paradigm-shattering. The first is that people weren't farming here, or anywhere near. The complex must have been built by hunter–gatherers. And the second is the date: Göbekli Tepe is 11,000 to 12,000 years old – 6,000 or 7,000 years older than Stonehenge (which is a fraction of the size of Göbekli Tepe).

Göbekli Tepe is supposed to be impossible. Here you have a huge monumental megalithic site, showing elaborate symbolism. To build something like this you need to inspire and co-ordinate a massive workforce. You need physical and sociological infrastructure. Göbekli is supposed to need, and to mark, an already stable, structured society, bound together by theological presumptions and strong leadership. Hunter–gatherers aren't supposed to need, want or be physically, organisationally or cognitively capable of something like Göbekli. But they were.

Sedentism isn't necessary for much of what we call civilisation or culture.

★

Stephen Mithen (whom we met when we were talking about music as a proto-language) wonders if domestication was a by-product of complexes like Göbekli. All those workers, revellers and worshippers would have needed a lot of feeding and a lot of beer. Those needs would have concentrated wonderfully the minds of the organising committee, and perhaps made proto-farmers ask themselves if the yields of wild cereals could be increased. And perhaps, Mithen speculates, some of the partying worshippers, hung over with einkorn beer and whatever other ecstasy was on offer in the forest of pillars at Göbekli, took some of the impressive new grains with them as they went home.

Wherever they came from, the new super-grains (robust seed heads and soft seed coats, springing up regularly each year rather than being long dormant, and without the long plumes designed to keep off the birds) found their way to the Jordan Valley.

The Natufians were waiting for them. And it's there (and most dramatically in Jericho) that we see true, unequivocal agricultural communities and burgeoning urban life for the first time. It is often said that Jericho was the first town. That may or may not be true. If it wasn't the first settlement to warrant that dubious honour, it wasn't far off.

★

The bus to Jericho goes from the Arab bus station near Damascus Gate in East Jerusalem, just below a grinning rock that credulous evangelicals think marks the site of the death and burial of Jesus. Soon it is rolling downhill along the road through the Judaean desert, past hilltop settlements (strategically crucial fortresses which just happen to have nursery schools, burger bars, herbaceous borders and populations of commuting IT professionals from New Jersey), past the occasional encampment of Bedouin with camels, goats and internet on their phones. Now, over the keffiyeh-wrapped heads in front of me, I can see the red hills of Jordan through the wobbling haze over the Dead Sea.

As the road flattens out onto the plain north of the sea, we are at the lowest point on earth. Most of the vehicles turn south along the west coast of the sea, past groves of date palms and the mountains where the scrolls of the Qumran community were found, and towards the bikini-lands of the Dead Sea resorts, where everyone goes to be photographed reading a newspaper as they float.

I'm going north, parallel to the last stretch of the River Jordan before it disgorges into the Dead Sea. We're stopped at an Israeli army checkpoint. Sometimes, to show the Palestinians that they can, they make everyone file off the bus and stand in the sun while they rummage through the little baskets of vegetables that the old women failed to sell at the Damascus Gate market, looking for the nuclear warheads and Russian tanks traditionally secreted under the zucchini. Today, though, the soldier boys are too bored to care. They just get sullenly onto the bus, look us insolently up and down, ask me charmlessly for my passport, flick cursorily through it and wave us on.

At the Palestinian Authority checkpoint outside Jericho

there's a different, more old-fashioned brand of militarism: handlebar moustaches and brisk epaulettes. They're even less interested than the Israelis, and a few minutes later I'm climbing out into Jericho.

I've been here dozens of times before, often as a refugee from the bitter winter cold of Jerusalem. This has always been a town of refugees, of the dispossessed – first those running from the dry cold of the Younger Dryas, and now Palestinian refugees who have fled the fever of the Arab–Israeli conflict. Its warmth and its water have historically been its big attractions, but now busloads of American fundamentalists in elastic-waisted trousers waddle through the souvenir shops en route to or from the supposed baptismal site of Jesus on the River Jordan, filling their nylon knapsacks with olive-wood camels because the olivewood nativity figures would look dangerously Catholic back home in Alabama. I first came here decades ago to sit in a restaurant, eat hummus, drink Turkish coffee, scribble in a notebook, watch the fights and hope that a Dutch girl I'd seen here once would turn up again.

Old Jericho – Tel es-Sultan – is an arid mound just outside the modern town, next to the Ein es-Sultan spring, which was the Natufians' – and one might argue the Neolithic's – main artery. In the first phase of permanent settlement (from around 11,500 years ago) Jericho was a small cluster of little round houses of bricks made from clay and straw.

I'm standing on the top of the Tel, in a hot wind coming out of Arabia that smells of oranges, burning tyres and cardamom. There's a swinging sea of banana trees in the plantation below, a bell ringing in the Monastery of the Temptation, grafted into the cliffs above, and minarets in the town are urging me, before it's too late, to surrender and pray to the One God, whose prophet is Mohammed. I share the Tel only with some thin dogs, who know exactly how far a stone can be accurately thrown, and stay away, though I'd like their company. Ravens are scouting along the crags, looking for a prophet to feed.

I can never stay here too long, because the thoughts pound relentlessly in like heavy surf. Could it really have been here that it all started? Isn't it too small to take the blame for factory farms, fur coats, chicken nuggets, GPS-controlled combine harvesters and the Keystone pipeline? Were cultivated grains first grown in that field, just down there, by the filling station? Did Natufian hunters, coming back from the pistachio woods with gazelles slung on poles, climb up to the Tel along the same path I've just taken?

No one knows where or when animals were first domesticated, but sheep and goats came first, followed fairly soon by cattle and pigs, and wherever and whenever it happened, domesticated animals were found fairly early here in Jericho. I imagine a tired, footsore hunter coming back to his wife in a little round house next to where I'm standing, and telling her that if he could only find a mate for that orphaned wild goat kid that she so kindly raised, they'd have effortless meat. And so, perhaps, it began.

And so, perhaps, it was here that our bodies started to become less robust than X's, and our brains started to shrink. (Domesticated animals, whether sheep, humans or, astonishingly, fish, have smaller brains than their wild counterparts. The reduction is particularly in the limbic system that governs awareness and general *aliveness*. Wild things – whether human or otherwise – simply get more out of their days and have more information about the world than things in hutches.) Perhaps it was here that sexual dimorphism, at least in sheep and goats, started to reduce: males no longer had to lock horns or appear desirable to attract a mate, since the females were handed to them on a plate. Perhaps it was here that the fertility of female sheep, goats and humans increased, and females started to be randier, to reach sexual maturity earlier and yet to be infantalised, keeping as adults some behavioural and anatomical features of juveniles that wild creatures lose when they reach reproductive age. Perhaps it was here that infectious diseases first started to flow from domesticated animals to the human population (most human infectious diseases are thought to have been transferred from non-humans). Perhaps it was here that infectious disease itself started to be a significant cause of death: bigger communities mean increased infection rates. Perhaps it was here that the occupational diseases associated with repetitive stress and unanatomical position were first seen: women who spend much of their day grinding corn with their legs curled under them get a characteristic form of arthritis in their toes, and many other types of arthritis are associated with agriculture. Perhaps it was here that dietary deficiencies first became an important problem. Hunter–gatherers generally had a varied diet, but monoculture means dietary vulnerability. Iron deficiency is seen for the first time in early agricultural communities. Perhaps it was here that tooth abscesses (from the grit in flour) and gum disease became common.

Perhaps it was here that boredom began. Hunter–gatherers had to solve a huge variety of seasonally changing problems, had used a huge variety of physical and cognitive tools and had to have encyclopaedic knowledge of many facts from many different domains in order to do so. In the new world there were fewer cognitive challenges: the difficulties were quantitative, commercial and, eventually, political – and such problems are trivial and tedious compared with the challenges of eking out a living and a thriving from the wild world. 'Can we produce enough of this food from those fields this season to feed our people?' they now asked themselves. 'Have we enough storage space for the surplus?' 'Are we better off trading with this village or that?' It was dull and pedestrian compared with hunting mammoths, retrieving souls from the underworld, knowing when there would be berries on a hillside fifty miles away, flying recreationally to distant stars on a chariot made of mushrooms and designing and building a new house every other night.

Perhaps it was here that leisure began to ebb away: where the idea of a holiday as a fortnight in the summer began. It's rare for hunter–gatherers to spend more than half their waking hours on the trail of calories. The agronomist Jack Harlan, using a flint sickle to cut wild einkorn in Turkey, famously showed that in three weeks a family could harvest the cereals it needed for the year.

Perhaps it was here that specialisation began: where one man became an expert in crop cultivation and another an expert in goat-raising – perhaps one becoming an expert in staying in one place and the other, a relative nomad, wandering over the thyme-scented hills after his flocks, sleeping out with them at night and keeping the old hunter–gatherer connection with the circling heavens. Perhaps, then, it was here that a wedge was driven between Cain the planter and Abel the pastoralist.

Perhaps, too, it was here that a wedge was driven between men and women. It is presumed from study of more modern hunter–gatherers that men and women always had different roles: women tended to gather, and men tended to hunt, but in most environments gathering was generally more important for survival, and that kept male egos in check. Now, in a cultivation economy, the women stayed at home and ground grains, and it was easier for men to boast that they, the men, were the primary producers and the women merely processors. The egalitarianism of hunter–gatherers has been overstated, but it is certainly true that on average they are a good deal less hierarchical than sedentary societies.

Perhaps it was here that the notion of surplus was born, and hence the notion of profit. Perhaps it was here that the population grew too high for shame and reputation to be the only necessary police, and so perhaps the first tyrants were here. Perhaps it was here that animals first started to be seen as things rather than fellow-travellers; here that the process of de-souling the non-human world began. Perhaps the idea of ownership was conceived here: of title and hence entitlement. Perhaps here, having partially de-souled the non-human world, the process of de-souling other humans started.

And perhaps from up on the mountain, where the monastery now stands, hunter–gatherers looked down pityingly on the farmers with the knowledge of superiority possessed by every crusty anarchist who looks at a merchant banker.

The American anthropologist James C. Scott has documented the attitude of hunter–gatherers towards agriculturalists. The executive summary is that agriculture is hard, boring work: not a proper job for hunters, and to be avoided if it can possibly be. Plough agriculture is particularly undesirable. The Europeans who colonised North America had to lock native Americans up in concentration camps to force their hands to the plough, just as the early Mesopotamian states had to rely on slave labour and coercion for their grain industry. When the European Black Death cut swathes through the population, leaving lots of spare land, plough agriculture was immediately abandoned for the older slash-and-burn model.

Today we are all farmers: we farm people (look at them in their open-plan, high-rise pens, pumping out profit), resources and, if only vicariously, pigs and chickens. We fear the haughty gaze of the hunter–gatherer; at some level we know that his contempt for us is justified, and we have done our best to rewrite history to say that hunter–gatherers gratefully embraced agriculture as soon as they could. It's not true. Scott has pointed out that the first evidence of sedentary communities comes from around 11,000 years ago. We see the first evidence of domesticated plants and animals at about the same time. Yet it is not for another *7,000 years* that there are settled villages, relying on domesticated plants or fixed fields. For 7,000 years, that is, our own model of human life, which we like to assume would have been irresistibly attractive to the poor benighted cavemen, was resisted or ignored, just as it is by more modern hunter–gatherers. Hunter–gatherers only become like us at the end of a whip. *Our life is a last resort for the creatures that we really are.*

Farming, like heroin, is easier to get into than out of. Surpluses boost population, and high population kills all the animals and eats all the nuts and berries from miles around, making return impossible. Once the jaws of a monoculture close around you, that's it: you've just got to go on producing more. And when you start trading, the law of supply and demand increases the pressure; binds you more tightly to the wheel.

You've taken the hard coats off seeds and altered the instincts of cattle. They can't survive in the wild any more than you can. You have to be around to defend them. Forget a year-long ivory-hunting or soul-searching break in the Derbyshire tundra: you've got to be on duty on your farm, protecting your crops and your stock from the vulnerabilities you've chosen for them. And if the envious people in the next village decide to beat their ploughshares into swords, you've nowhere to run to (as a hunter–gatherer you had the whole world as a refuge) and no resources, either physical or imaginative, to survive anywhere other than the farmstead. And anyway you couldn't take all those little babies the corn has given you. Perhaps Jericho was where we lost all our options.

What a charge to level against this quiet, warm place with its trinket shops and orange groves.

The sun is dropping into the wilderness to the west, the lights are going on in the town and it is time to get on a bus that will shudder up the hill back to Jerusalem, where I'm staying in a Crusader cellar.

We pass the Bedouin again. They're tending their camels, herding their goats and changing a wheel on their Toyota pick-up. They might be pretty compromised examples of their kind, but there's no doubt that the God of the Judaeo-Christian tradition prefers them to the field-farmers of Jericho. Yahweh romanticises itinerant pastoralists far more than Rousseau did. These Bedouin are the descendants of Abel, whose name means something like 'vanishing breath': here one moment, gone the next, as it is burned off by the sun rising over the mountains of Judaea. You can't have any delusions of grandeur if you're called 'Vanishing Breath'. You'd never be at the helm of a publicly quoted company, or have a condominium overlooking Central Park, or a well-stocked humidor. But you'd see the spring flowers that your brother Cain (the root of whose name is about acquisition and possession) drives past fast on his way to the board-room or the estate agency, and know the names of the birds without wanting to sell them off.

The story of these two brothers encodes and expounds the age-old motif of the settler and the wanderer, the cultivator and the pastoralist. The wandering pastoralists of the Near East aren't Upper Palaeolithic hunter–gatherers – this part of the Bible joins the story after the Neolithic is well consolidated – but Abel is far closer to being a hunter–gatherer than his brother. Cain is a 'tiller of the ground', like the ones I've just seen in Jericho, and Abel is a 'keeper of sheep', but without the pick-up. They both bring offerings to God: Abel the fat portions of the firstlings of his flock, and

Cain the 'fruit of the ground'. They are not equally pleasing: 'the Lord had regard for Abel and his offering, but for Cain and his offering he had no regard.'

Why does God prefer Abel's offering? It's not clear. Many elaborate explanations have been offered, and none of them really works. It looks to me like a simple, personal preference for Abel's character and lifestyle.

Cain is petulant and angry because Abel's offerings have been better received. They go out 'into the field', and Cain, like his neo-liberal descendants, snuffs out the competition. He kills Abel. The very first human violence occurs on the quintessential Neolithic territory – the field.

God passes the sentence that Cain had been dreading most. He'd have to sell the condo, abandon his share options and for ever wander the earth like the brother he had killed. He'd become like the homeless people, huddled under their blankets, that he walked past so smugly on the way to the office. His roots would be cut.

But the strangest part of the story is that, until you look closely, the sentence seems never to have been served. It looks as if it's going to be. Cain goes to the land of Nod – which literally means the 'land of wandering'. But he doesn't wander there. He *'settled'* there. He went straight back to the bad old ways that had resulted in the sentence. And not only that: he built a city, Enoch, named after his eldest son. It's a classic illustration of the Neolithic's obsession with kinship and lineage: an attempt to beat off death by wooing posterity. The city flourished, in so far as any city can. Population soared, industry throbbed (Tubal-cain made 'all kinds of bronze and iron tools') and no doubt its products were traded, and there was a vibrant metropolitan culture (Jubal was 'the ancestor of all those who play the lyre and pipe', the Bible says, forgetting about all those Upper Palaeolithic bone flutes).

Had God forgotten the sentence? Had Cain got away scot-free? No. In the end we're all sentenced to get what we want. Cain wanted urban stasis, the delusion of security: pension policies, cunning investments, a big house with an electrically operated garage door and the pick of the shopping malls. He got them, poor bastard, and that was the sentence. He could have had the flight of the eagle, the sun on his face, the laughter of his home-schooled children. He could have starved for a week, spun round a fire of acacia wood, become a gazelle and dived into the springs at Ein Gedi and been fed by ravens in the mountains of Moab.

It might not seem like it, but Abel Vanishing-Breath, despite – or rather because of – his vanishing breath, won.

But there is one more piece in this part of the Judaeo-Christian (and, now, Islamic) jigsaw.

The bus is now coughing through Arab East Jerusalem. Spread out and sparkling below us, the golden Dome of the Rock at its spiritual centre, is the city holy to Jews, Christians and Muslims. 'Next year in Jerusalem,' Jews say every year, Christian pilgrims pant up the medieval Via Dolorosa, believing it to be the route taken to Calvary, and Mohammed came to the Temple Mount on his winged horse, al-Buraq, establishing a long tradition of Muslim veneration of the city. But it's a *city*. Quite definitely, with five-star hotels, a sewage system and ATMs. If God is being consistent, shouldn't he have said that the Passover prayer should be something along the lines of 'Next year in some desolate wilderness, where you can commune with nature and get a real sense of your place in the natural order'? And isn't it embarrassing that for the Christians, the New Jerusalem that rises at the end of time is also a city: a shining, sparkling, deeply unnatural one, all crystal and mirrors?

Quite the opposite. The Passover and the Revelation of St John are redemption stories. They're saying that if *even cities* can be redeemed, there's hope for us all, even if we have to wait until the end of our lives or the end of the world for it. God's bias to the wanderer has been clear from the beginning, and it hasn't changed. The identity of the Hebrews was forged during those forty years of dust and blisters in Sinai, and when they finally settle in Jerusalem, what is at the very centre of the holiest place in the holy city? It is the Ark of the Covenant, in which God was carried from campsite to campsite. It never loses the poles that allow the portable, Bedouin God to be lugged around. It is always a mobile shrine for an itinerant people. Black goatskin tents are the natural habitat of Islam, and the institution of the Hajj reminds Muslims that there is merit in movement. And, finally, that very Jewish Jew, Jesus of Nazareth, seen by Christians as the ideal human, observed that while foxes have holes, and birds have nests, the Son of Man has nowhere to lay his head.

In short, according to any of the religions that grew up here, where farming started, you're gambling scarily with your soul if you take up either actual or metaphorical farming.

★

I'm generalising too wildly. I'm judging the whole by the worst. I have tendencies that way.

I can think of many places to go for a better perspective, but one of the best is Fran and Kevin's place, in a remote part of mid-Wales. They don't just talk about the Neolithic; they live it. They farm the flighty little mouflon that are the ancestors of modern sheep; they have erected standing stones; they make and bake their own clay bowls, weave their own baskets, drink home brew from horn beakers, make fox-skin hats, tan skins in brains to make them soft and in boiled oak bark if softness isn't so important; and they're going to be buried in tombs visible from miles away on the hill – 'so we can keep an eye on things'. And though they themselves live in a medieval farmhouse, they have built a small settlement in the British Neolithic pattern: round houses with wooden lattice walls packed with mud, thatched with heather, a single low entrance and a mud floor.

This is no pose. They are deadly serious. They live this way because they have decided, very carefully, studiously and reflectively, that this is the best way to be alive as a human being. And make no mistake: these are most splendid, most alive humans.

Despite Fran's detailed directions, Tom and I get hopelessly lost, and the midwinter sun is near the crest of the hill by the time we trundle along the woodland lane to the farm. A tame wolf yaps at our wheels, and as soon as we go into the house Fran hands us a bowl of domesticated aurochs meat which has been bubbling alongside a pot of beeswax, charcoal and pine resin glue which is used for fixing flint axe heads into wooden shafts.

'Eating this cow is hard,' says Fran. 'It's always hard to kill things you've loved. I feel stressed for days before the killing. I can see why Neolithic people needed to have a feast to look forward to, to numb the pain. It was much harder for them to kill animals. The animals were much closer to them, and much harder to raise. The animals were also their insurance policy: the reserve that could make the difference between life and death. You don't lightly rip up a policy like that.'

For Fran and Kevin the slaughtering and the butchering are a community venture. The steak, the tripe and the guilt need to be shared. Their feast isn't a gluttonous death party, but rather an acknowledgment that the killing of an animal is a morally serious act. It needs laborious justification, and if many people get pleasure from it, the justification is easier. I agree. We only have meat on high days and holidays, when my agonised utilitarian calculation concludes that the net amount of happiness in the world will be increased by the life, death and consumption of the animal. Fran and Kevin are more practical. 'We shoot rabbits to spare the sheep for as long as we can.' And of course, as Tom and I felt with the hare, it's a moral imperative to use absolutely everything from the animal: not just the meat and the offal but bones for tools, skins for clothes, tendons for thread.

Tom and I are in one of the round houses tonight. Within minutes we've trodden things into the floor that will rewrite the history of humanity and wreck academic careers when they're excavated in 10,000 years' time.

It's a clear night, and bitingly cold ('like a wolf's nip,' says Tom). The constellations are busy tonight, harrying one another. We make a heather fire, feed it with birch and coil dough round sticks to bake in the embers.

X and his son are back, squatting nearer than ever, looking covetously at our lush down jackets and thick mittens, puzzled at the snakes of burned dough that slither into our mouths. They must be camped in the wood just below our settlement: there's nowhere for them here.

I get a good look at the boy for the first time. He's stringy and distracted, with none of his father's self-possession. Long black hair falls low over his face. Only the fire and Tom can keep his attention for any time. His eyes come back to the fire to rest, as if it's home, and to Tom, as if he is a lesson he's been set. When he tugs them away from the flames or Tom they dart all around, and his lips press tightly together. His father sometimes looks long and hard at him, but the boy never returns his gaze.

A mile or so away a calf is stuck on the way out of its mother, and by the tempo of her bellowing the cow is getting tired. On a hilltop farm behind us a dog is straining at the chain, desperate to get at a fox. A badger pushes through the bracken, head down like a snow plough, forcing the stems apart. A satellite slides down Orion's leg.

Tom says a sleepy goodnight and goes into the hut carrying a burning branch to light the way. I hear him struggle into his bag, and then he slings the branch out of the door. A few minutes later he's snoring gently. His head will be pushed determinedly into the heather he insists on using as a pillow, and thrown back, unwinding the adolescent hunch he adopts in the day.

It is hard when a child goes to bed. It makes you feel grown up, which isn't good. It is hard to stare across the fire at X and Co. I am disoriented. If the Neolithic arrived in Britain around 6,000 years ago, and X was self-realising in Derbyshire 40,000 years ago, 34,000 years separate them – that's nearly six times as long as the distance between me and the start of the Neolithic. Some time during those 34,000 years, humans acquired the drive to mastery, and a new era began.

I'm doing it again – libelling a whole massive epoch. Perhaps they didn't acquire the drive to mastery here. Neolithic Wales, just like Upper Palaeolithic Derbyshire, was an edge place. It was difficult to live here. It still is. To survive here then you had to co-operate with the natural world, not seek vainly to enslave it. You had to be a jack of all trades, just as the old hunter–gatherers were. To be monocultural was to be dead. I suspect the old ways of relating to the heavens, the earth and the dead survived for a long time here. The geography doesn't lend itself to creeds or priestly dynasties.

It's time for me to sleep too, but when it's this cold I've always found it hard to sleep – not because of the cold itself, but because you've got to cocoon yourself and pull a blanket over your face, and that's like being dead. We can't really light a fire inside the hut. There's no smoke hole. Smoke filters eventually through the thatch, but not before it has filled your eyes and your lungs. I don't like it that there's no smoke hole. I can see the stars through the open door when I push my head out from under the blanket, but it's not the same. I want very much to be able to lie on my back and see the heavenly hunter and his hounds rampaging through the night. Close the smoke hole of someone's yurt, say the Siberian nomads, and you've excommunicated him, cut him off from whatever divine presence there is, locked him up in his cell to be raped by his own psyche.

It's about three in the morning, I guess. (I've haven't worn a watch or had a phone for years.) Tom's snoring isn't just making me maudlin; it's making me mad. I might as well go out, and while I'm out there I might have a leisurely crap, watched only by the owls.

The effort to get up, shod and out in the cold dark is like no other comparable act of will. I am a weak man, and it takes me half an hour to summon the strength of character.

Once I'm out there, I wonder why I ever went to bed. It's always the same. I never learn, and never remember. 'What a deal man misses out of life by lying abed', wrote the great nature writer 'BB'. I've missed a hunt, and perhaps a kill, somewhere near the Pleiades, and a hunt, and a certain kill, down by the barbed wire fence, where the blood is black in the light of the moon. I've missed the sheep processing out of the next field and lining up neatly like guardsmen around the fire, and the little breaths of the birches joining to become a cloud making the badgers cough and grounding the lower-altitude owls.

When Neolithic farmer–hunters left their huts to defecate, they must have seen their situation and their bodies differently for a while. This was the only truly lonesome activity. Everything else – birth, feeding, sexual intercourse and death – was communal. And sheep were still sufficiently people for lone shepherding to be communal too. Only the squatting body was really alone, and only when looking back at the huts from the squat can they have seen them as they were, and started to piece together a sociology.

★

We're only here a couple of days, extravagant though Fran and Kevin's hospitality is. We can't be tourists in the Neolithic – visiting other people's projects, immersing ourselves in them for a while, getting a taste and then coming back to make notes. For the whole thing about this era is *responsibility* – to a place, to crops, animals and humans. The responsibilities that started then are still the ones that keep us on the treadmill, whether onerously or not: rents, taxes, marriages, bedtime stories. But the land is a far more ruthless tax collector than a government. It demands not only sweat, dung, money and time but also fidelity of thought and moral purity. For in the Neolithic mind, and in the mind of most farmers that have ever been, right-living and right-thinking are rewarded and dark deeds and twisted thoughts are punished. An adulterous act or an unauthorised killing with a stone knife makes the crop fail.

Do you want to be authentically Neolithic? Being a decent citizen, you probably are already. If not, patch up the garden fence, eat chops from a captive pig, check the deeds of your property, flick sentimentally through your family photos and worry about dying.

SPRING

'Death, be not proud, though some have called thee
Mighty and dreadful, for thou art not so;
For those whom thou think'st thou dost overthrow
Die not, poor Death, nor yet canst thou kill me.'
John Donne, 'Death be not Proud'

It has been a winter without stories, mainly because I thought I'd understood the Neolithic story, and the story about the Neolithic that I'd adopted was political. Big, wrong stories always suffocate all other stories, and since all political stories are wrong, all politics diminishes the colour, complexity and downright entertainment value of the world. No political story can tell anything that is really true about any human. I feel hollow and smutty when I'm engaged in political discourse of any kind. All politics defames all humans.

My dead father wasn't there at all throughout my Neolithic winter. He was an abiding memory, of course, always at my shoulder; but he was more of a set of principles, or a pair of disembodied eyes to whom I would ultimately be accountable, than a body in an armchair, smelling of coal tar soap and pipe tobacco.

His absence should have alerted me to the inadequacy of my exploration of the Neolithic, because the whole of the Neolithic, from start to finish, would have been diagnosed by modern psychiatrists as suffering from Persistent Complex Bereavement Disorder. If you think my constant reference to my own father is mawkish and unhealthy, you're a long way from your prehistoric roots. The writer Julia Blackburn ate her dead husband's ashes with yoghurt. That's what normal humans do.

I've written a good deal about my father in this book. I've barely mentioned my mother. She died too, as is the way of things, but before my (older) father, which is not generally the way of things, on a glorious spring morning with her beloved daffodils blooming in the garden, as is often the way of things in this taunting, tantalising universe that speaks of resurrection just as it turns out the light.

She was a schoolteacher, a dazzling musician and linguist, and a very free spirit. So on her death my sister and I set about, with the help of a tame undertaker and a dodgy solicitor, carrying out her funerary wishes. After the doctor had come, murmured some kind platitudes and filled out the death certificate, we stripped her naked, carried her now stiff body out into the garden, laid it on a trestle table that my father used for putting paste on wallpaper, and I sharpened a brand new set of carving knives that I'd bought in the market.

'Do we really have to do this?' I asked my sister.

'It's what she'd have wanted,' my sister replied. She's always been the upright one. 'You go first: you're the vet.'

The first cut was the hardest. I thought I'd start with a thigh rather than an arm (she'd hugged us with those) or a body cavity. As the knife went in, and I recognised the layers (skin, subcutaneous fat – not much of that – fascia, muscle) I was able to be abstract about what I was doing. This wasn't particularly *her* muscle: it was just muscle. In fact it wasn't really her at all. She had gone, and we could discuss later where she'd gone to. After that, it was hardly jolly, but my sister joined in with a will, and in an hour of hard work we'd skinned the body, defleshed the limbs, and removed the viscera. We piled the muscle and the various organs onto the compost heap, lit a fire with the help of a lot of paraffin and soon the cul-de-sac smelt of the biggest barbecue it had ever known.

That, it turned out, was the easy bit. Jointing the body was a nightmare. Although she hated walking (unless it was round an art gallery), her ligaments were like steel, and before we could break her up and fold her into bin liners for the journey to the family vault, we were sweating.

All this, of course, is pure fiction. As you read it, you were disgusted and thought I was a deviant who should be locked away. But why?

The Neolithic was the time, *par excellence*, of the ancestors.

In the first phase of Jericho's permanent settlement some, at least, of the dead were buried under the floors of the houses. You'd walk, cook, squabble, teach and have sex on top of your parents. In our Oxford kitchen there are photos of our dead mothers and fathers. Even those anaemic memorials inhibit some of our more extravagant cruelties and grossnesses, and inspire some of the very few acts of nobility to be found in our house. It would be powerful indeed to be able to appeal to the children, in their internecine wars, on behalf of the grandparents a few inches under their socks.

A skull cult developed in Jericho, in which the features of deceased relatives would be modelled on the de-fleshed skulls, the eyes replaced by shells and the whole no doubt displayed in the house. That would improve the homework no end.

In early Neolithic Britain there were corporate tombs, containing many generations, often built on the top of, or adopting the plan of, actual houses of the living, and earthworks (known as causeway enclosures) consisting of concentric ditches, in which large quantities of human, and sometimes animal, bones are often found. When you died, you would literally, in the words of the Bible, be gathered to your ancestors; your bones mixed with theirs. Eventually.

What often happened in the tombs is that newly dead bodies would be left to decompose on top of the more experienced corpses, or were stacked at one end of the chamber, so that having moved from the estate of the living to the estate of the dead, the journey of transformation continued incrementally, with maggots, rats and beetles as the pilots. The tomb was a tunnel, like a stone vagina, along which the dead passed on the way to rebirth. Self-transformation didn't finish with biological death.

The tombs – and particularly the megalithic chambered tombs of western Britain – were sometimes busy, social places. They have forecourts and other spaces to accommodate living visitors. You'd take a picnic. They were meeting points – not just for the living to meet one another for consolation and to show and feel solidarity in their shared grief, but also for the living to meet with the dead: to show respect and to keep the ancestors on side, to get the instructions of the dead and to invite their continued participation in everyday life. No doubt the bones would be stroked, kissed and handed to the young children who had never met their dead grandfather. Now they had.

Death must seem much less final, dramatic and frightening if you know that you'll be visited and held for ever more, and that your relevance to your family will *increase* when your breathing stops.

Sometimes the tombs themselves were meat processing plants where the bodies were stripped of their flesh with flint knives (which have left their marks on some of the bones to this day). Sometimes the corpses were processed elsewhere (perhaps at the causeway enclosures) and brought, already de-fleshed, to the tombs. Often bones were rearranged once they were flesh-free. Skulls were sometimes placed around the edges of the corridors and the chambers, and at the Lanhill cairn in Wiltshire long bones were stacked between two rows of skulls, with a fully articulated skeleton barring the way to the door, presumably awaiting treatment that never arrived. Bones were evidently moved from site to site, sometimes leaving bits behind. In the West Kennet long barrow, near Avebury, there aren't enough skulls, femurs or tibias. The smaller bones of the hands and feet must have been missed when the rest of the bones were moved from another place.

The landscape of early Neolithic Britain was generously littered with human bone. 'It is no overstatement', writes the archaeologist Julian Thomas, 'to say that in earlier Neolithic Britain the remains of the dead were ubiquitous.' And not just in tombs, causeway enclosures, ditches next to long barrows, isolated pits, caves and rivers but in bags, homes and, if they had them, pockets. I carry a string of Greek *komboloi* – worry beads – around in my pocket. I fiddle with them constantly. If I'm walking, or sitting in a café, I take them out and swing them around and clack them together. In the Neolithic I'd be doing that with human phalanges.

But human bones weren't just toys. 'We might talk,' says Thomas, 'of a general economy of human remains.'

The small bones of the hands and feet are hard, and more resistant to cremation. After we burned my father (I hope, but doubt, that the calories from a lifetime of steak and kidney pudding were recycled to heat the local primary school), we were handed his ashes in the brass-cornered box and discovered a number of finger bones, blackened but firm.

My father was always very fond of my friend Burt, who farms in the Welsh Black Mountains, and his wife, Meg, who is a witch. And so when we next saw them, I gave them a couple of the bones.

'Look after them, won't you?' said I. And Burt, with a solemnity I'd never seen in him before, promised me that he would, and hugged me, and so did Meg, and they put the bones high up on a shelf, next to a stuffed gull, where the children wouldn't find them. And there they remain.

Burt and Meg's family are closer to us than ever; closer now than blood brotherhood (another relationship founded on an exchange of biological substances), and when their parents die, we're getting some of them for our mantelpiece.

'It's a strange thing,' says Meg. 'When we're arguing, we sometimes look over to the shelf, and your dad seems to sort it out.'

Again, all untrue, I'm afraid. But that's how it would have worked in the Neolithic. Thomas compares the circulation of bones through the Neolithic with the circulation of objects in a gift economy. The exchange of bones created and cemented relationships, not just between the givers and receivers but also, perhaps, between the living and the dead. The dead still presided over the dinner table; still negotiated contracts, made matches and judged law suits.

It seems to me supremely logical to take your ancestors' bones (if not their mummified bodies) with you as you go about your business – probably in the early Neolithic, walking across miles of country after the domestic animals. Isn't that, after all, rather like what I was doing by taking round with me the leaves, cones and pine needles my father picked for me? The dead want to continue to speak, to be felt, to influence. If their bones aren't pressing into your neck as you lie down to sleep, perhaps they will use other channels. Perhaps the lingering scent of the coal tar is in lieu of the charred metacarpals that we meekly buried in a Somerset churchyard. The dead help us to be ourselves, just as our parents in life, by DNA and by example, made us what we are. If you are defined (as you are) by your ancestors and by your journeying, you can't really be yourself as you wander if the ancestors are left at home. Quite apart from anything else, it's a grave discourtesy and unkindness to leave your parents in a wet hole. It offends the proprieties of the prehistoric world, as would most of our conduct. X, although he doesn't think about his ancestors in the same way as these Neolithic herdsmen, would no doubt be even more outraged by the sight of a municipal cemetery on the other side of the bypass as he would have been by the sight, smell and sound of Steve the Peedo's abattoir. Neither is the way that life and death (if they're different) should be done.

It's important not to be anachronistic about the idea of personhood, though. Thomas wisely cautions about the danger of a straightforward equation between Neolithic bodies and 'individuals' in the contemporary sense.

> [T]he notion of a person enclosed within one skin and containing a soul or a mind is a curiously modern Western one [...] Archaeology, as a practice of modernity, materialises ancient bodies through a medico-scientific mode of understanding. We need to recognise that this is quite remote from the ways in which those bodies would have been lived [...] [M]odern Western conceptions of personal identity and bodily integrity probably did not apply at the time.

Indeed. In the Upper Palaeolithic, though there was a finely developed sense of self, it was a self whose skin was permeable to the whole world, human and non-human. To say where 'I' was would have required the use of an infinite number of cross-bearings, taken from the position of rocks, flowers, wolves, wives and stars. In the early Neolithic the number of cross-bearings thought to be necessary reduced massively, to the members of the human community and to the domesticated animals around whose lives the human community revolved. (Cattle skulls have been found mingled with human bones in several long barrows, suggesting, according to Julian Thomas, 'some form of equivalence between human and cattle remains'.) And as the Neolithic wore on, the number and variety of self-defining, self-locating relationships reduced still further.

The most obvious sign of this in the British archaeological record is the shift from long communal barrow tombs to discrete graves (some of them marked by individual round mounds), a very gradual process which Thomas believes started at the end of the fourth millennium BCE. Round mounds are characteristic of the early Bronze Age (from around the start of the second millennium BCE), but there are Neolithic round mounds too, and some of the later long barrows cover pit graves containing articulated bodies.

The dead of the *later* Neolithic were far more emphatically and individually dead than previously dead humans had been. Inhumation in a mound was much more of an endpoint than being stacked and shuffled in a long barrow. It wasn't just the recently dead of the later Neolithic who were barricaded away out of the light, away from the living; it was the ancient dead who lay in the old chambers. In the west, where the chambers had been designed to facilitate conversation between the quick and the dead, the entrance to the chambers was often blocked with stone and earth, which muted that conversation, and these efforts were sometimes accompanied by attempts to bring dispersed or disarticulated bones back together. The dead now had their place. So did the living. And now the addresses were different.

This consolidation of bones located the dead much more definitely in the landscape. In the early Neolithic, if you asked where your grandfather was, you might be told: 'Well, some of his ribs are up on the causeway enclosure, his right femur is with your aunt and his left with your uncle, I have his pelvis, some of his feet are rattling round at the bottom of the bag hanging up by the wild boar, a dog had his spleen, there's an ulna in the river, a humerus by the blasted elm, a heel bone in a hole by the old oak tree, and the rest of him is in the chambered tomb where we go on Sunday afternoons.' If you asked the same question in the later Neolithic, the answer would be a finger pointing to a mound, and the word 'There'.

The early Neolithic landscape was inhabited generically by the dead. Discrete points in the later Neolithic landscape were occupied by particular dead people.

This sounds political: it was. It goes to the root of self-perception. If your bones stay together, there's a clear dead 'you', as there isn't if you're scattered across the chalk hills of southern England. If you're going to be discrete when you're dead, you will tend to think of yourself as discrete when you are alive. And if you can point to a place where your ancestors are, you are going to be much readier to talk about your entitlement to that place and the surrounding area, and to regard the ancestors who are buried there as the justification for your claim. That's the instinct at work in Rupert Brooke's famous sentimental poem 'The Soldier': 'there's some corner of a foreign field/ That is for ever England.'

The round mounds of the later Neolithic and the early Bronze Age weren't as lonely as all that. Other bodies might join you; cremated remains might pepper the mound; and the original mound might be the seed of a cemetery. But everything would depend on, would refer back to, would be justified by, the original burial. Just, in fact, as the legitimacy of a claim depends on the legitimacy of the original possession. The geography of a late Neolithic cemetery looks very like a modern diagram showing familial relationships. Self-definition became a matter of genealogy. Remember those interminable Old Testament chronologies? A begat B, and B begat C, and C begat D. That's the mantra of the later Neolithic. Since one could be certain about one's ancestry ('Look, you can *see* my ancestors: they are those lumps on the hillside') one could be certain about one's rights to the land. *Tenure* burst into Britain, and became consolidated in the field systems that persisted at least until the Romans arrived.

Earlier Neolithic people had had a much more fluid relationship with the land. They had settlements, but for much of the year they were wandering pastoralists, and their dead were everywhere. There must have been territorial claims, but much of the landscape was left fallow for much of the year, and there was little reason to get aggressively protective about most places of interest to the herds and flocks. Much of the land was heavily forested, and if a herder wanted to clear some forest to graze his stock more effectively, he could often take what he wanted: there were plenty of flint axes and fire-spitting stones.

Walking and the absence of walking proved, again, to be politically momentous.

I want to believe that with the idea of tenure came an intimate, loving, stewardship relationship with the 'owned' piece of land – a relationship closer and kinder than that of the relationship between wanderers and the land over which they wandered. But it doesn't look like that to me. Clutching anything makes you want to clutch more, and clutchers don't tend to be kind.

★

'Well, that's nice,' says Meg. 'Thanks a lot.'

Meg and Burt (who turned out not to have my dad's metacarpals on their kitchen shelf) farm sheep, cattle, trees, heritage grains and children in the Black Mountains of Wales. They gave up raising pigs because they liked them too much, and they make their own electricity and socks. They live on vegetables, roadkill and chorizo made from sun-dried DEFRA officials. We're sitting outside their farmhouse, drinking home-brew, eating radishes and looking over the river towards an Iron Age hill fort that Meg runs up every morning. They're exhausted. It's lambing time, they've been up all night and out all day, and they smell of amniotic fluid and K-Y jelly.

We're a few pints down, and I've been ranting about acquisitiveness and the relationship between the birth of farming and the birth of avarice. They take it far too graciously, but when I start on about the notion of dominion being connected with farming, they lose it. Very gently, though. They just start laughing, loudly and uncontrollably.

'You're joking, right?' splutters Burt.

'He's not,' says Meg, and they collapse again, helpless. And I hear myself saying again how the early farmers and growers were the first Nietzscheans; jackbooted annexers of territory whose languages they had long forgotten; self-appointed barons and, soon, self-appointed gods; ruthless, covetous, high-handed, uninhibited even by reflections on their own transience because of their belief in their own immortality, or at least the immortality of their dynasties.

'Now listen,' says Meg, suddenly serious. She puts down her glass, which always means trouble.

'We're terrified, right? All farmers are and always have been. Mastery? Don't be fucking ridiculous. If the weather turns, all our barley will be gone. If foot-and-mouth disease comes to visit, we'll watch as generations of animals, bred by six generations of Burt's family, are shot and burned in a pit. Any moment that fucking mountain' (she gestures at the hill fort) 'could slide onto us. And you're telling me we think we're serenely in control? That we really think that we run this place rather than it running us?'

Burt is looking at his boots.

'You're a fascist, you know that? You think that everyone apart from your beloved hunter–gatherers in your fucking Golden Age, who understood everything about everything and lived in perfect harmony with everyone and everything, is part of the great herd of unrealised *Untermenschen*, sliding ever deeper into the moral swamp.'

She has a point. She always does.

The debate about the significance of the Neolithic is, like most debates, dangerously polarised. Polarisation is always a sign of intellectual laziness, and I'm more guilty of it than most. Farming, on a small scale, can be a way of enacting, not trashing, the hunter–gatherer ethos. Big-scale farming, though, is almost always disastrous, whatever one's measure of disastrousness is. Many of the ecological, political and psychological woes of England can be laid at the door of the enclosures, which snatched common land, incorporated it into larger private farms and cut the umbilical cord with the land that fed the rural rank and file and had infused wildness into the English psyche. I'm about to admit all this, but Meg hasn't finished.

'Why are we here, do you think? For money? Don't make me laugh. We'd get more by doing a paper round. Because we like the view? A view to you is a factory floor to us. I'll tell you why we're here.'

She takes a long swig of her beer, and pours out a stiff tot of sloe vodka.

'We're here, though I squirm to say it, because we *love* the place. Love it although, and possibly because (psychoanalyse it as much as you like), it's constantly threatening to ruin us and kill us. Maybe it's a kind of self-love: I grant you that. It certainly has got inside us and we've got inside it, and I don't know where I end and the hill begins.'

Burt looks up from his boots.

'There we are then,' he says. And indeed we are. Meg's is the voice of the Upper Palaeolithic, and there indeed, leaning on the tractor, smiling, nodding and sniffing a placenta that the dogs had missed, is X.

★

It's about four o'clock on a warm spring morning. I've just woken up. I'm half-sitting, half-lying, propped up on a straw bale in a lambing shed looking onto the broken land of a Scottish hill. A ewe, conceived and impregnated on that hill, is blowing sweet hay-breath – last summer's crystallised sun – into my face, but that's not why I'm awake. I'm awake because of a throaty moan that might mean a problem.

It does. The ewe must have been straining for a while. I should have seen and heard it before. I slip a hand inside and groan. The first lamb has the worst of all possible presentations. It's what's called a dog-sitter. All I can feel is the middle of the back. The head and all the legs are pointing back into the uterus.

I know what I'm supposed to do, but I'm useless at these tricky manoeuvres. My hands are too big, too clumsy, and I'm dangerously impatient. I flounder around for a while, trying to straighten out the hind limbs so that I can pull the lamb out backwards, but there's very little room in there, and I'm worried about damaging the uterus, the lamb or both. It's time to give up, so, embarrassed and humiliated, I hammer on the farmhouse door.

Janice, the farmer's wife, comes down, smiling, as if I were the mid-morning postman with an anxiously awaited parcel. Nothing along of the lines of 'You're the bloody vet', or 'Couldn't you have left her until after breakfast?' The lamb is out in ten minutes, followed by another. Soon they are head-butting the udder, then they're attached, and then Janice and I sit on the bale and watch.

'It's wonderful, isn't it?' says Janice. 'I've seen it thousands of times, but it's always special. It's always the first time.' She gets up, dusts herself down and heads off to make the porridge.

This is a commercial farm. That lamb will have a number, a place in an electronic ledger, and will soon find its way to the slaughterhouse and onto a dinner plate. Janice will account for the tax and VAT generated by its body.

Perhaps, in thinking about the Neolithic, I am seeing only the corpse and the ledger, and failing to take Janice's wonder into account.

Much of my own understanding of otherness comes from the eyes of Scottish sheep and Derbyshire cows. The first time (and one of the only times) I felt useful was shovelling cow shit in a Peak District farm when I was ten. It had a dignity that piano lessons, cub scouts, arithmetic and even amateur taxidermy did not. That wasn't because I thought I was helping to put milk on Sheffield's cornflakes. That never occurred to me. What I was detecting was that humans acquire their significance from relationship, that relationships with non-humans are vital, and that clearing up someone's dung is a good way of establishing relationships.

★

Over the years I have often slept among farm animals – in lambing sheds in the spring, on hay in a cowshed or out with the cows at pasture. Not with pigs yet, though I'd like that.

The restlessness and watchfulness of cows is abidingly fascinating. I can attend to their attention for long absorbed hours, I tell Burt. Cows know the dark, the light and the field so much better than I ever can. I want to know their field

– or know anywhere at all – as well as they do. If I knew anywhere – even a piece of grass the size of the handkerchief with which I mop the trousers onto which I've just poured my beer – as well as that, I would know something about the way the world is made that I've wholly failed to sense, let alone grasp. But there's more. It seems to me, as I drink more and more of Burt's beer, that the field needs their lowing, cudding attention to continue to exist.

I stand up unsteadily, raise my tankard to the hill fort and declaim Ronald Knox's parodies of the philosophy of Bishop Berkeley:

There was a young man who said 'God
Must find it exceedingly odd,
If he finds that this tree
Continues to be
When there's no one about in the Quad.'

Knox recorded God's response, and I bellowed that out too:

'Young man, your astonishment's odd.
I am always about in the Quad.
And that's why this tree
Will continue to be,
Since observed by, Yours faithfully, God.'

That's how, at this point in the evening, I feel about cows and fields.

'Cows, Burt, are like God: necessary watchers.'

I might even be serious: it's hard to know. Without their attention, Burt, my dear, dear friend, the field would evaporate. The idea that the world needs a beholder wasn't an eighteenth-century innovation: it's ancient and ubiquitous. It's only polite, my old cock. Don't you tell the kids to look at the person they're talking to, who in turn will look back at them? Of course you do. So do I. Well, then: if the human and non-human worlds are packed with persons, they're watching you, and they need to be watched back, or else they will turn away.

Alan Garner, drawing on this tradition, writes in *Boneland* (I go on) about how Alderley Edge has to be seen at all times by the anointed watcher: '"I have to be able to see the Edge from wherever I am", said Colin, "in order to keep it. If something isn't looked at it may go, or change, or never be."'

'So, then,' I go on, 'if there's no dedicated Colin, and you've burned and felled the trees and eaten the wildlife, the cows are the only possible watchers. If that cow turned her face from the corner of the field, it'd be gone, and where would you be then, eh?'

Burt tops up the tankard.

'If those were my cows, and I tended them as I know you do, I'd feel responsible for the field enduring: I'd feel that I was watching and so maintaining the world through the cows. And, bizarre though it sounds, I reckon that Neolithic farmers justified like that what's crudely and inaccurately glossed as their "ownership" of land. How about that, then?'

'You're howling mad,' says Burt. 'And you're trying too hard. You should get out more. Drink up and have another.'

★

Meg can't tell where she ends and the hill and the sheep begins. This is a sort of rustic Welsh *advaita*: the non-dualism sought after by spiritual searchers of all traditions, and notably by those of the East. But there are limits: she's very good at drawing a line between her and her sheep when the sheep are loaded onto the trailer and driven off to have their throats slit. And that's the big divide between the Upper Palaeolithic and the Neolithic, however early.

The Neolithic learned to draw lines. There are no straight lines in the natural world, of which Upper Palaeolithic humans were a part; no clear divisions even between species. The Upper Palaeolithic prayer to the animal killed or about to be killed was: 'Let me, exceptionally, do this without the reciprocation that is the usual rule. Let me eat you without being eaten myself – at least for the time being. I know my time will come, because I'm basically like you.' Farmers never pray in quite that way to their own animals. They couldn't. It would make farming psychologically unbearable. There has to be a 'them' and there has to be an 'us', and the two have to have a different moral status. This was the basis of the Neolithic. If one can talk meaningfully about a Neolithic revolution, it consists in the recognition and enactment of us-ness and them-ness. There had been a sense of self before, but now it was radically reconfigured, and cast primarily in negative terms: 'What am I?' asked Neolithic humans. 'Not one of *them*,' came back the answer, setting the scene for many horrors.

The lines between things don't exist. Nor do lines themselves, until we make them. They are only in our heads: in our mental worlds. Yet the actual, historical Neolithic is criss-crossed by lines. We had started to remake the actual world in the image of our self-created one. We had started to destroy the world and replace it with models of ourselves.

★

House martins are snapping at rising flies over a lake. There are so many dragonflies that the view across the reeds is fuzzy with interference. If we had proper ears, we'd be deafened by the crunch of chitin in scissor jaws. Tom and I are in the Somerset Levels, not far from Glastonbury, on the Sweet Track, a straight two-kilometre Neolithic walkway between a ridge and what, when it was built in 3807 BCE, was an island. It was originally made from oak planks, lifted above the marsh on oak, ash and lime pegs driven into the soggy ground. We're walking on a sturdy modern replacement that for a while takes the old course.

It was part of a network of similar raised roads which gave access to islands that held stands of reeds for thatching and weaving, animals to spear and plants to pluck. The marsh had said: 'No through road.' Humans said: 'We write the rules now.' When writing about the Neolithic, it's impossible to avoid the language of rape, which is also the language of colonialism and pith-helmeted exploration. The wetlands were penetrated. Stakes were rammed into the peat. The islands were taken, subdued, conquered, ravished.

Rape is done with a straight, erect penis. Representations of erect shamanic penises are just about the only straight lines you see in Upper Palaeolithic art, which is generally as curvaceous as the rump of a bison or the hips of a woman. The only other common straight line in Upper Palaeolithic art is seen in the mysterious ladder-like symbols found on cave walls, but there they are combined to form non-linear shapes. Hunter–gatherers have straight artefacts, but they generally pierce body cavities.

The Sweet Track is the only straight thing here. We're so used to straight things that we don't notice the anomaly. It's not lost on X and his son, trailing behind us, looking in stern wonder at the line of the walkway, mesmerised. For that's what straight lines do: even to us. They take your eyes and your mind captive. They trammel vision: they shrink the land. I'm watching Tom, and it's happening to him. He's looking straight ahead. Three dimensions have shrunk to one. A sparrowhawk has just killed a greenfinch to our left.

He didn't notice. A swallow miscalculates and grazes the skin of the water with a wing tip. It wasn't straight ahead, and so he missed it. Lines take eyes – and so thoughts – captive.

Captivity is another context in which lines first emerge in the Neolithic. The Neolithic saw the first fences to contain animals: to stop them being themselves; to shut down their experiences; to make them conveniently available. Fences cut up the land, leaving wounds, making it less of a whole. The undivided land had been nameless; had been itself. The parcels of land all had human names on them.

Back in the car we spread out a modern large-scale map to trace the route of the Sweet Track. Tom's not seen a map like this before. He points at it.

'What are those lines?'

They are the lines showing the fields.

'Why are they so straight?'

'Because someone decided they should be. You see the same thing on loads of maps. When we get home, look at the map of North Africa. Someone just sat in a room and said that the borders should be there and there and there.'

'I bet the swifts don't care where the borders are when they're flying here from Africa.'

Indeed they don't. But human lines have reshaped the lives of many birds.

We go to Glastonbury to eat our sandwiches, sitting on a bank and watching the house martins. Their crops may be full of insects harvested that morning from the Sweet Track pool. They curve and circle and make parabolic swoops, for that is the way of things, but then their eyes are drawn, as Tom's were, to a straight line: the line of the eaves where they nest. Martins' eyes have been drawn to these lines for many thousands of years now. Before humans started building houses the martins nested in caves and on cliffs, but now they are as addicted to straight lines as we are.

It's not just humans who make strange choices: we facilitate the making of strange choices by other creatures. These Glastonbury martins could have chosen to raise their young on an Umbrian cliff, feeding on the rafts of food that float up from the olives, listening to warm wind and nightingales. It would have been shrewd. The little balls of mud they use to build their nests would have stuck more firmly to the rock than they do to the plastered breeze-blocks of the Glastonbury housing estate, and you get a better class of ballooning spider there. The insects here are choking on diesel fumes and pesticides, and the nestlings have to listen to Heart FM.

Humans have had straight-edged buildings for a long time. The appearance of rectilinear structures correlates almost perfectly with the appearance of agriculture. Correlation is not causation, but there is a strong case that agriculture, and the world view that it embodied and promoted, straightened the curves of the Upper Palaeolithic.

The general picture is clear: hunter–gatherers throughout Europe, south-west Asia, the Near East – and indeed more widely – tended to occupy circular buildings. With permanent settlement and agriculture came rectilinear structures. The typical transition is seen clearly in south-west Asia: in the Pre-Pottery Neolithic A (*c*.10,000–8800 BCE), where there was a shifting, field-less lifestyle, there are only circular dwellings. In the next phase, the Pre-Pottery Neolithic B (*c*.8800–6500 BCE), when agriculture as we recognise it is seen, the dwellings are generally rectilinear. From then on, at least for buildings associated with the everyday business of life, rectilinear buildings are the norm.

(Britain and Ireland, as so often, are extremely odd archaeologically. When farmers first arrived in Britain they had rectilinear longhouses. But then they bucked the trend, and went back to circular buildings. This is the dramatic exception that proves the rule.)

Our houses often expound, sometimes savagely, our views of the world. Ours is cluttered with brawling children, skulls, icons, beekeeping equipment, musical and surgical instruments, bottles of formalin with voles, eyes and embryos dimly visible through the sediment, like a demonic caricature of those snowstorm bottles that you shake, unclassified and unclassifiable books, stuffed auks with awkward wings, rocks in which I could once see a face and hope I might again, hopeful and hopeless seedlings, and faded prayer flags, all awash with cider. My father, grey-bearded and corduroyed, looks lovingly from a Tuscan hill towards my mother, who, youthful and flowery in a meadow near Bath, is holding the score of *Rigoletto*, and the smell of coal tar soap moves unpredictably between rooms. You would know us reasonably well if you could bear to stay here for a week and your immune system and your nerve held out.

But had we exercised any real choice about the shape and location of our house, you would know some far more fundamental things about us: things we might not know ourselves.

Some of the choices we make might look functional. But that's rarely the whole story. Circular houses, for instance, resist high winds rather better than rectilinear ones, and so, if an angular house is to survive anywhere windy, it has to be aligned carefully. But another way of putting that is that a circular house (and therefore its inhabitants) *belongs* to the system that includes the winds, and an angular house does not. The very word 'angular' is eloquent. Rectilinear houses stick their elbows into the natural world, push other things out of the way as they decide what position they should have. To decide on the fundamental orientation of one's living space, and hence one's life, is a declaration of independence from the forces of nature.

Circular houses are hard to extend, whereas a room can

easily be added to a rectilinear structure. Inherent in that simple observation are the ideas of both permanence (and hence tenure) and progress. Rectilinear houses are there to stay, because this is my damn house on my damn land, and we're going nowhere, and the house is going to annex even more land and get bigger and better because that's the sort of people we are: we're going places, even though we're staying right here.

A circular house is an intrinsically democratic space. At the centre is the hearth, rather than a person, and the central hearth gives out its heat and light equally to everyone because that's how the physical laws of radiation work. A circular space has to be shared with others: there is free flow around it. It's hard to have secrets in a circular one-roomed house.

Rectilinear structures are very different. There's your room and there's my room; mine is bigger and more richly furnished than yours, and in my room I do things about which you know nothing. There I hoard things, my precious, and in the heat of my torrid lonely nights make plans to impoverish and supplant you.

Houses may model the owners' cosmos. Ours does, which terrifies me. The hunter–gatherer has a huge, overarching sky that reaches down to the earth on all sides, as do the walls of a Mongolian yurt or a Natufian round house. A farmhouse is mission-focused and hence self-focused, set up to survey the pasture or the grain store, pointing directionally and metaphorically to the important things in life, which happen to be owned and at least notionally controlled by the farmer. Farms and houses, like lines, tunnel vision and constrict the inhabited world.

But I am ignoring a crucial class of building: buildings for the dead and for the gods. The general rule – to which there are some exceptions, such as passage tombs – is that from the Neolithic to the early Middle Ages people who ate, slept, rutted, schemed and reared children in rectilinear buildings used circular structures (tombs and stone circles) to mark the non-quotidian, and rectilinear buildings were often surrounded by circular enclosures, as if to indicate that the ruling metanarrative differed from the one that dictated everyday conduct.

What should we make of this? Whole professional lifetimes are spent arguing about the meaning of stone circles and other megalithic monuments, but most agree that stone circles and temples are concerned in some way with the dead, and that they are places of power. To say much more is to play with fire. The stones in a stone circle may represent individual ancestors, or the circle may represent the whole community of the dead.

The French archaeologist Jacques Cauvin, who specialised in the prehistory of the Near East, suggests (credibly to my eye) that, universally, circles represent the transcendent and the whole, and that accordingly they are often a proxy for femininity, fertility and intuitive understanding. Rectilinear forms, on the other hand, represent the apparent, immediate, concrete world – the world of the masculine. This, if right, would make stone circles places in which one had access to a direct, intuitive type of knowledge: a knowledge in which the ancestors had graduated, and which could be channelled to the living. It would also make them important in the mediation of fertility. Death and fertility were surely linked inextricably in the minds of the early farmers. They knew that unless the seed was buried like their grandparents, there would be no new life. Prayers for harvest must have been high on the list of the petitions to the powerful ancestors.

This, like most attempts to see into prehistory, is speculation. But perhaps a closer look at the later Neolithic monument of Stonehenge will take things a little further.

For many of us, Stonehenge *is* the Neolithic. We think we know it: elephant-grey blocks crouching on Salisbury Plain, besieged by tour buses or ringed by white-robed druids. But it is very, very odd. There's nothing very like it anywhere else. A standard Neolithic building project it most certainly is not. Yet, as in all scholarship, as in all life, the outliers are often the most eloquent.

It's a bad mistake to try to generalise too much, and a criminally naïve mistake to generalise too much on the basis of the evidence from highly atypical Neolithic Britain, but Stonehenge may be unusual simply because it is much more explicit than other monuments about the relationship between the living and the dead.

The monument we think of when we think of Stonehenge – the huge dressed pillars and lintels, morticed using the techniques of domestic carpentry, and the smaller blue stones, brought 150 miles from Pembrokeshire – used as a cemetery from the third millennium BCE, is just one part of a massive complex. It has a wooden counterpoint nearby, originally made of wooden pillars and probably unroofed, and now known, conveniently if not imaginatively, as Woodhenge. A few metres away from Woodhenge is a Neolithic village, Durrington Walls, which had its own timber circle, and probably housed Stonehenge's builders.

Today Stonehenge is usually full of tourists throughout the year, eating ice cream and talking about mistletoe and human sacrifice. But when it was properly in use it was frequented only occasionally – probably particularly at the winter and summer solstices.

No one but the dead stayed at Stonehenge itself. But people did come to Durrington Walls, at huge expense and inconvenience, and there they feasted mightily, and no doubt drank and fornicated copiously. Falstaffian quantities of pig bones have been found there. The atmosphere must have been rather like the Glastonbury Festival, but it's one thing to drive from Fulham to Glastonbury for the weekend in your cruise-controlled Range Rover, and quite another to walk the whole extended family, babes on your back, driving your herd of pigs before you, from Yorkshire to Wiltshire. That's what some of them did.

Probably they came to lose the fear of death, and to recruit the dead to their cause. Other purposes have been suggested, such as healing and the unification of disparate communities. All are possible. Doubtless the lame and the blind asked the powerful ancestors for help, and even if unification wasn't the intention, it would inevitably have resulted from such big gatherings.

The Woodhenge–Stonehenge complex must be seen as a whole, and perhaps the best way to see it is as a metaphysical theme park showing the juxtaposition of life and death.

Woodhenge represented transience. Other wooden henges were deliberately burned down or dug up and removed to make their transience more obvious than natural decay would. Woodhenge's work was to represent the current life of the revellers, and it did so at least partly by withering. Now you're standing upright like these pillars, said Woodhenge, but look a little closer (at your feet, if you're a middle-aged man, or your varicose veins if you're a woman), and you'll see the signs of rot, just as, if you look more closely at the pillars, you'll see beetle-holes.

But the pilgrim-revellers didn't stay at Woodhenge, brooding on their mortality. They walked. Just down the River Avon from Woodhenge there's an avenue. It is straight for much of its course, and it leads to Stonehenge.

What is death like? Stonehenge and Woodhenge together answered the question, and St Paul gave the same answer 1,300 years later: the perishable will become imperishable, the rotten immovably solid.

Life and death, like Woodhenge and Stonehenge, were linked by a straight line. The line could be paced out in the theme park: the journey we all take could be rehearsed and so drained of some of its terror. If pilgrims walked metaphorically between the two realms, they wouldn't be so scared when they had to do it for real. The land made comprehensible the connection between life and death: pilgrim feet could feel the connection. The *land*, unlike death, was not a realm of inchoate mystery: not a place of dark and smoke, of which poets could only hint vaguely. Having walked the way the pilgrims, unlike St Paul, didn't see through a glass, darkly. They saw clearly and confidently. It was all there in the land. They'd walked it. Perhaps we see the legacy of that confidence when the Romans invaded and were intimidated by the sight of a people who appeared not to fear death.

They did not fear it, for they knew it was not the end. Seeds went into the earth, and new life sprang from them. The dark would never win.

The Woodhenge–Stonehenge avenue was the manifesto of a new intellectual and spiritual imperialism. Not only could humans burn forests and wall up the wild; they could comprehend the metaphysical mechanics of life and death. They ruled not just sheep but eternity. It was a big claim.

There are two other Neolithic lines to note at Stonehenge. First, there are its walls. They don't look like walls to us: there are big gaps. But the dead couldn't squeeze through the gaps. Those mighty sarsen stones are there to keep the dead in their place. Stonehenge is a prison. The length of the avenue and the sweep of the river separate them from the rollicking of Durrington Walls.

In the good old days of the early Neolithic, penning up the dead in one place would have been insufferable apartheid. They knew then that the living and the dead needed one another. As we have seen, the bones of the dead were everywhere, dead voices contributed to every conversation, and the dead often had the casting vote in important decisions. Unless you could smell coal tar soap, or see a hunter–gatherer and his son out of the corner of your eye, the world wasn't working properly.

The second line is the line of time itself, drawn in the late Neolithic in the course of drawing lines of descent. If you claim a field because your ancestors farmed it, you'll make sure that you can establish direct, *linear* relationships with the past. Those relationships were celebrated in the ancestor cult at Stonehenge.

This was a wholly new way of seeing time. Time had previously been cyclical. The seasons had come and gone and come and gone, and the ancestors were always there to help you through the seasons. It made possible the previously unthinkable notion of *progress*, which in turn generated a whole new cult of status and denigration.

★

Tom and I, watchless, clockless, pointless, planless, mapless, are meandering across a hot hill in the Spanish Sierra Nevada, the cooked soil and stewed vegetableness of Africa in our noses from across the mirror sea to remind us where humans are really from. Sometimes the wind spins round and sheep bells confound the fantasy, but mainly it's eagles at the edge of sight, the skirr of distant waves, and scree that was washed out of the slope 40,000 years ago and might have been last disturbed by the broad feet of a Neanderthal hunting party, or last Wednesday by the shepherd racing to get home in time to see the Manchester United match.

The first and only fields up here were the fields of the dead, which belonged, like the old, bone-studded causewayed enclosures of Neolithic England, to no one living.

They were therefore neutral, safe places for negotiation and fiesta. Death will level us all. That fact should transform our politics, and in fact, for a while, it determined the shape of Neolithic politics. The dead have discovered, as we all will, that status and possession are ludicrous. On the turf of the dead the living have to play by the egalitarian rules of the dead. It is likely that causeway enclosures and similar charnel grounds were indeed used for communal deliberation and dispute resolution. It was good practical psychology. We all behave better in cemeteries; we feel watched and judged by our forebears, and the vanity of human ambition is inescapable. A mass grave is the only proper place for a parliament or an international summit conference.

It is high here. The air is thin. Tom is finding it easier than I am. Like a dog off the lead, he walks five times further than I do, scouting in the spiky wax-leaved bushes for grasshoppers, pouring water into a crow's skull to measure the size of its brain, using the shiny black back of a beetle as a reflector to make light-spots creep up an oak branch, whistling down a hole in the hope that the snake will whistle back.

But tonight we will be back in a whitewashed mountain village from which the Catholics brutally expelled the Moors, eating slices of meat hacked from a dried pig's leg hanging from the ceiling, sitting next to a fire constrained by stone walls and forced to breathe up and down a brick trachea.

'*La li-li-li, li-li.*'

SUMMER

'[T]he domestication of animals and the breeding of herds [...] developed a hitherto unsuspected source of wealth and created entirely new relations.'
Friedrich Engels, *Origin of the Family, Private Property and the State*

Imagine finding a pretty baby wolf.

The children like her, so you decide to keep her.

If she grows to her full size, you realise, she'll be too big for the house and will need long, exhausting walks, which you'd never manage. The obvious solution is to starve her so that she's stunted, and smash her legs.

That works well. The legs heal in an interesting bowed shape. That gives you another idea.

Her nose, which was stubby and cute when she was a cub, has elongated. She now looks rather wolf-like in the face, which is a bit sinister. So you take her out to the garage, turn up the music to drown out the whining, pick up a claw hammer, and smash her face in.

There's a lot of blood and a lot of fuss, but it soon heals up, and the result is great. She's got a very attractive squashed face. She can't breathe so well, of course. She wheezes badly, but you get used to that, and it has its advantages. It means that she doesn't want to go far or fast on the walks.

What you have, of course, is a pug. X and his son saw one

as they walked behind me and Tom along the Oxford towpath one dull summer afternoon. I thought they were going to put a flint-headed spear through the owner's Barbour. They didn't understand that the bone and face-breaking were genetic and that the owner's guilt was, well, corporate.

I share that guilt. I don't eat many animals, and I try to eat only ex-happy animals, but I'm cavalier, and no doubt over the last few months I've eaten a chicken whose breast muscle was so big and its legs so weak that it couldn't stand (not that it would have had anywhere to walk to if it could), beef from a Belgian Blue that had a genetic mutation resulting in massive proliferation of its muscle fibres, with consequent difficulties (often needing Caesarean section) in squeezing out the huge-buttocked calf, and meat from a sheep that should have had its mother's uterus all to itself but had to share it, in the name of profit, with two others, causing a lot of maternal pain – and no doubt foetal trauma too – at delivery.

I've drunk milk too. A modern dairy cow in a conventional unit might produce 30 litres of milk a day, with 4 per cent fat, 3.2 per cent protein, and 4.8 per cent lactose. That's 1.2 kg of fat, 1 kg of protein, and 1.5 kg of lactose. You simply can't wring that out of grass, or indeed out of any cow that eats normal food at a normal rate. You have to feed the cow food that will be fermented faster and passes faster through the rumen, leaving room for the next mouthful. It comes at a cost: the cow has reduced fertility and immunity, and a higher risk of rumen acidosis and lameness. She spends all her productive life on a metabolic knife-edge. We get tasteless milk that's more of a dietary liability than a benefit, life-endangering antibiotic resistance and rivers poisoned with the slurry.

Control, design, hubris, hidden and accumulating costs. Which continue to accumulate today. The Neolithic.

There are still plenty of *early* Neolithic farmers around,

though. Their ruling dead are everywhere – not just under the local family mound – calling them ethically to account. They have a liability to the past: to the bequest they have received. Their responsibility, they say, is to everyone, not just to their own genetic successors: not just to the people who will eventually join them in the mound. They wander the world, figuratively if not actually, though they come back to the rectilinear farmhouse (which they have no intention of extending) at night. Ask them what they are, and they'll say they're part of an ecosystem – an ecosystem that's not just to do with composting, the recycling of animal and human dung, the preservation of invertebrates, the installation of bat boxes and the smiling of sheep, but also with the primary school fundraising quiz, the co-operative that runs the local bookshop and café, and the cooking of dinners for the demented.

But they are an endangered species. The late Neolithic is hounding them out of business.

I went to meet a late Neolithic farmer. I'll call him Giles. He's the friend of a friend of a friend, and he greeted me smoothly and cautiously when I went to his vast Lincolnshire farm. It hummed with motors, not bees. His small, soft handshake was the only small thing here. He was nattily dressed, with a dazzling white shirt and corn-coloured trousers.

He showed me into the farm office, which was hung with framed certificates and photos of machines, and waved me to a chair.

'So, what is it you want to know? Fire away.'

What I really want to know is not the kind of knowledge you can get by firing away, and so I said, vaguely, that I'd be interested in some basic information about the farm: how much they produce, what their challenges and aspirations are and (more daringly) why they did what they did. He relaxed at this. That was something he could do very well.

'Wheat,' he said. 'We do wheat. Big time. Fifteen hundred acres. All very scientific. We got thirteen tonnes per hectare last year – that's about five and a quarter tonnes an acre in real money. I'm a 700 wheat heads per square metre man, which, though I say it myself, isn't bad. And why do we do it? The world has to be fed, doesn't it? Four and half billion people eat wheat every day. We're doing our bit.'

'So,' I asked, looking out of the window at the pancake-flat fields, 'are you master of all you survey?'

'I am,' he said contentedly, leaning back in his chair and joining me in the view. 'I am indeed.'

He drove me round the farm in a brand-new vehicle with heated leather seats.

'I don't apologise for that,' he said, when I commented on the seats. 'Important to be comfortable when you're living your life outdoors. I do the same for my chaps. Only the best for them. Entertainment too in their cabs: best system money can buy. Can get tedious out there.'

I asked him how many workers he had.

'Two,' he said. 'And first-rate they are too.'

Two men in 1,500 acres. Population density wasn't much different in the last Ice Age.

The drive took a while. Every field looked the same to me: ripening wheat, each stalk the same height, each head drooping with profit. But they weren't the same to Giles. That one was planted with a different strain, one that needed slightly less nitrogen than its neighbour.

'We spend thousands on soil testing. Got to get it right.'

I'd heard about the impoverishment of these eastern soils. In the past, crop rotation, fallow periods (a variant of the old Jubilee idea of giving the land and its inhabitants rest and freedom once in a while) and the dung of grazing animals would have kept them in good condition – which meant, among other things, a vibrant microbiology and a texture that would stop them being washed and blown into the North Sea. Now, it's said, there aren't many harvests left.

'Nonsense,' said Giles. 'Alarmist liberal poppycock. My grandsons and great-grandsons will be farming here and making a good living from it; don't you worry.' He looked meaningfully at me over the automatic gearbox. 'The Lord won't give up on this land, or on my family. He's given us dominion over the land, and He won't withdraw the commission. We've looked after the land, and He'll look after it too, and after us.'

Giles, it turned out, was an old-school conservative evangelical Christian. The local Anglican church, like the Green movement, was too liberal to stomach. 'They soft-pedal sin. They don't tell us how we really are. And most are, I'm afraid,' he went on happily, 'destined for an eternity of torment.' His sons, away at the moment at boarding-school, shared his convictions, he was glad to say – which is why he could be so confident that the Lord would continue to smile on his balance sheets from generation unto generation. The highlight of their year, apparently, was going to a Christian camp for other boarding-school pupils, where they were taught that humans were intrinsically bad and that it was more important to evangelise than to be kind.

That's my cynical gloss, I'm afraid, drawn from a leaflet that Giles thrust into my hand as I left, along with a fact sheet about the volatility of grain prices.

Most of the farmers I know – whether early or late Neolithic – are religious, and the ones who insist most loudly that they're not are the ones who insist most loudly that they are 'spiritual'. Neolithic farmers were certainly religious and – if it's different – spiritual too. I've suggested that Neolithic religion had its origin in the farmers' need to wash the blood of ensouled but non-human cousins from their hands. Once the new Neolithic religion had ignited, its flames were fanned by the winds of terrible contingency.

Hunter–gatherer life was relatively safe. If the caribou didn't come, the band could move elsewhere to live on salmon. But agriculture changed things: it was the business of putting all one's eggs into one basket. If the grain failed, that was it.

For hunter–gatherers there were many more contingencies, but none was likely to be life-threatening. And what was more, they could negotiate directly with the contingencies: they could beg the caribou to come back, or the lightning to hold off. The farmers had fewer contingencies: they needed sun, rain, the absence of rain for harvest and the absence of disease, but any one could wipe them out. And to whom should their petitions be directed? It could only be to a god or gods up there – not in the visible world – who sent or withheld all the necessary blessings. Once the direction of prayer was established as vertical, priestly hierarchies, regulating the flow of intercession, became inevitable, and social status and power began to correspond to access to the priest.

It's not so different today for many farmers. Meg's right: farmers live all the time under the shadow of terrible possibilities. For most of us that dark shadow would mean perpetual and pathological gloom. The weather forecast might mean ruin rather than just a damp walk to the bus. If a man in a suit in Whitehall has an off day, or a supermarket buyer thinks the parsnips aren't symmetrical, they might lose their market and their house.

The red-faced farmer in the tweed suit who hands round the collection plate on a Sunday morning prays much more earnestly than the rest of us. He must. If you're a slave to a particular bit of land, you need also to become a slave to a deity.

These seem to be the rules:

Few but serious contingencies + insight = Sky gods
Many but negotiable contingencies + insight = Animism
　　or its ilk
[Few but serious contingencies OR Many and serious
　　contingencies] + No insight OR [Insight + Huge
　　courage] = Modern humanism

★

I'm lying on my back in a meadow of long grasses. Grasses,
not grass, because there are many species of grass here, and
many non-grass plant species among them – perhaps 150
grass and flower species in all, winding, creeping, prickling,
each of their heads swinging to a different tune in the breeze
from the sea. On my chest I can count at least eight species
of insects, and there are dozens more making the air throb
around my head. I'm hopeless on birdsong, but even I can
pick out fifteen or so.

Fran, back in the winter in Wales, had tried to convince
me that Neolithic clearances had increased biodiversity. I half
believed her. Dense forests had covered Mesolithic Britain,
and the floor of dense woodland is often fairly sterile because
little light gets through. I can see that letting in light by
hacking or burning would create other niches.

This meadow, part of my friend Kirsty's Yorkshire farm,
made Fran's point beautifully. It also made beautifully the
point that it's what the Neolithic *led to* that's the problem. I'd
hear to what it led if I climbed the fence onto the farm next
door. Or rather, I *wouldn't* hear. It is silent there. Its peace is
complete: it is the peace of the grave, because it is a grave.
The insects have been poisoned. There's nothing for birds
to eat. The only crop is wheat, doused with herbicides and
pesticides.

The traditional meadows, whose seeds were sown when

humans started managing the land, enhanced the natural world. Biodiversity is a good measure of ecological health, and using that metric, humans made the world healthier for a while. It's just that we never know when to stop. Perhaps part of the Neolithic syndrome is that we just can't.

Another important question is why we started. We've noted the reluctance of hunter–gatherers to leave their lifestyle and take up farming. Why didn't they just say no, and carry on as they were? I mentioned the North American hunter–gatherer communities who, cycling seasonally through political and sociological systems, knew what government and authority tasted like. Why didn't they opt for freedom and tax-free ease?

There must be many reasons. It's easy enough to understand why the Natufians who settled in Jericho and nearby regions wanted, literally, to put down roots after the traumas of the preceding few years. Those years hadn't been pleasant or representative examples of the hunter–gatherer lifestyle, but they had gone on long enough for memories of the old way of life to fade. The unusual conditions of the Younger Dryas had the effect of defaming hunting and gathering, souring them in the folklore of the Near East and giving sedentism a halo it really didn't deserve.

I knew an old Indian man who had come to England as a teenager. His family house in the Punjab had been burned to the ground by a mob in the tumult of Partition. He expressed his gratitude to England by being more English than the English. I never saw him without a dark suit or a tweed jacket. Instead of comfortable collarless shirts he always wore a starched collar and a *faux*-regimental tie. He read smug, dreary leaders in *The Times* when he could have been reading the Upanishads. He ate fish and chips rather than the ambrosial vegetarian curries of his youth, and duly died of a heart attack. As he looked out into the Yorkshire drizzle he saw only sun. He never missed an opportunity to dismiss Indian culture as 'primitive' and praise the courtesy of the English, and was deaf to the relentless patronising and racism of his neighbours. He would never have gone back to India. After a while he *could* not have returned. He is a model for the early permanent settlers in Jericho.

My parents grew up in the Second World War, when food was scarce. After the war, foul but novel food became freely available: all the processed foods that should make us gag. They were so grateful for the new abundance, malignant though it was, that for the rest of their lives they were addicted to processed food, preferring it over fresh. They too are models of early permanent sedentism in the Jordan Valley.

None of this is to pass any kind of judgement on my parents, that old Indian man or those early Near Eastern farmers. Their individual choices were not only comprehensible but rational. No doubt I'd have done the same. The Neolithic revolution rumbles on in all of us.

Once the first generation was entrenched, the die was cast. You need a vastly greater suite of skills to be a hunter–gatherer than a farmer, just as you need a vastly greater suite of skills to be a farmer than to work on an assembly line putting the same screw into every car. The skills necessary for hunter–gatherer survival, let alone thriving, were soon lost.

The farmers bred fast: the hunter–gatherers were demographically out-gunned, and the species they needed to survive were killed to ease the tedium of a cereal and goat diet. The farmers could begin to write the rules for all humans, not just themselves. One modern African group, previously foragers, explained that, against their will, they converted to cattle pastoralism because cattle had become the standard currency in the area. The bride-price had to be paid in cattle. No cattle, no women, no future. That's how it has worked from the start of sedentism. Hunter–gatherers were over the barrel in the Jordan Valley.

These are the sorts of thing you can say in archaeology and anthropology books, but there is something else going on too. We quickly lose not only the ability to be free but also the desire to be free. That desire has been almost completely extinguished today. Given a choice between air-conditioned slavery with a regular income and happy, penniless anarchy in a shed looking at the mountains, almost all of us, without hesitation, opt for the slavery.

At some level we know it's a bad choice, and we hate being reminded of the choice. Cain knows not only that he's less happy than Abel, but also that Abel is his superior. When Abel, leading his wolf on a string, dances across Cain's path as Cain's on the way to the office, Cain senses Abel's natural aristocracy and it makes him mad. And so Cain tries to destroy Abel, herding him and his family into concentration camps, requiring him to have an ID card and a passport to stop the wandering that Cain so fears and envies, and firing tear gas at him at Occupy demonstrations.

★

Symbolising took us over. We symbolised ourselves to death, abolishing all real things and valuing only the representations of them that we made. That terminal phase came long after the Neolithic, and in the final part of this book we'll visit the University of Oxford and eat, drink and talk with people who spend many hours every day pretending that representations are the real thing, or better than the real thing. In the Neolithic it wasn't nearly so bad. But nonetheless a sheep stopped being fully a sheep, and became instead a meal, or a profit, or a management problem, or a reason for self-congratulation – all of which, whatever their merits, are not sheep.

I think I'm immune, of course.

I'm no farmer, but I've played at farming over the years, and now I have a hand in various community agricultural projects. One's a city farm where we've got sheep, chickens, sometimes pigs and lots of vegetation. Another's a space where we grow vegetables in a good old Tolstoyan commune, leaving it to everyone to decide how many pumpkins, courgettes, peas or potatoes that afternoon's sweat is worth. And another's a farm perched up above the ring road, looking over towards the daydreaming spires where all those earth-shattering, earth-denying abstractions are hatched. We've got a lot of cows there that will end up as burgers, and some goats, because we like goats. Once in a while I go to one of these places and listen to a sheep's chest, pick up a foot, wring a bird's neck, dig potatoes, mend a polytunnel, sew up a cut, pass the death sentence and drink tea.

Now I'm up at the cattle farm. It's high summer, and the grass is high. Every day the cows get moved on to a strip of new grazing, and the electric fence is moved to enclose the new grazing area. My job today is to cut a new corridor for the fence with my scythe so that the grass doesn't touch the wire and drain the current.

This is a very Neolithic job. Sickles for the cutting of grass (cereals are just grass, of course) are the classic, defining Neolithic tools; the point of the job is to control the living space of another species: I control by cutting a dead straight line through a wild-ish environment, and then use technology to stop the other species from occupying its chosen world instead of the world I have chosen for it.

I love it. It is one of the most sensual pleasures there is. First I unsheathe the scythe from its canvas scabbard and test the edge with my thumb. I fill the whetstone's holster with water, hook it on my belt and dip the whetstone. I sweep the stone up and down the blade. 'Zing! Zing! Zing!' This is the most swashbuckling action in my life. It's how pirates start their day.

Next I mark out the line in my mind's eye. Today I'll start at that blackberry bush at the field's edge, and go straight up to the hawthorn. It's hot and blue, and mid-afternoon, and I'm in a T-shirt and shorts. It'd be better to cut in the early morning, when the grass is still wet with dew, but I couldn't manage that today. I've got a big jug of cider, and I drink from that before I get started. I'll sweat it out faster than I drink it. West Country harvesters used to be part-paid in cider: 'Agricultural lubricant', one Somerset cider maker calls it.

I pick up the scythe. Unlike the lumpen English scythes, which take the skin off your hands and make your discs pop out like coins in a fruit machine, it's light, perfectly balanced and fits me like one of my own arms. It's Austrian, designed for cutting hay in alpine pastures to the sound of cowbells. It does most of the work; I just have to swing it over the ground, swivelling my torso on my hips like a tank turret, with light pressure from my right arm. Get good at this and you can work all day. Scything was often regarded as light work – work for women, children and the elderly. It's fast, too, if you're good. A competent scyther will beat a petrol-driven strimmer any time.

I eye the grass that I'll cut with my first swing. Why has that been chosen to fall today, I wonder, through long and agonising habit, before realising that one answer, however incomplete, is that that's what *I* have decided.

You never know quite how the grass is going to be. I cut a strip yesterday, at about the same time, but today might be completely different. I take a deep breath, and start to swing up the hill. 'Swish! Swish! Swish!' It is beautiful to see the stalks topple as the blade, a crescent moon in carbon steel, slices into the pale, tender underbelly near the ground. Today

I can feel the crunch of the silicon in the stems: yesterday was slicing marshmallow.

I'm dispossessing thousands of insects. I break rhythm as a rabbit runs between my legs, and pause to drink cider, but nothing else stops me. Slash into the grass as I breathe out, back as I breathe in: 'Swish! Swish! Swish!' The breath, the cider, the sun and the sap are working, and I have been doing this for ever and will never stop and my boots are green and part of the field and X's head is peering at me out of the hawthorn and nodding suggestively at the cider jug. 'Drink up!' I shout to him, and he wades through the grass, looks hard at me with his brown, smoke-reddened eyes, bares his teeth in greeting so that I can see scraps of meat trailing out of the gaps, and takes a halitotic swig, and then another.

My dad waves at me from the top of the hill. He's sitting on a log, as he always used to do. He's wearing the trousers he bought in Brocklehurst's in Bakewell, and the stiff old boots he tried to give to me, and a check shirt washed, for some reason, in coal tar soap. Now that I know he's here, he looks away into the distance, trying as he always did to understand how the far distance can possibly be connected to the here and now. I wish he'd come down and have some cider. He'd find out who X was, would bombard him with questions as he always bombarded the friends who came to parties, or who got a lift in the back of our car. He had everyone's life story out of them in minutes, and the kids went away thinking he was really weird, and the adults went away feeling significant.

'Swish! Swish!' Flies are drowning in the sweat that's pouring off me. Their thrashing legs make little fluttering pulses against me. More cider, and for X too, but my dad must have found a breeze up there, and he's Olympian in his detachment – if any of the Immortals ever wore a broad-brimmed tweed trilby.

The cut grass is folded neatly down. At the bottom of the field it's already well on the way to being hay, and the dispossessed already have new communities. To my left, though, they shouldn't get too comfortable. This time tomorrow they'll be fermenting inside a rumen. The ones on my right, who'll be on the safe side of the fence, will have to wait a while for that.

'Swish! Swish! Swig! Swig!' Nearly there now. X has gone back to the hawthorn tree, and it's packed with sparrows, jostling for space around him.

Five more swings: 'Swish! Swish! Swish! Swish! Swish!' And I'm down on my back, the fluted columns of a grass cathedral soaring up past me towards the blue, a carrion crow wondering if I'm about to die and the kids, just arrived, shouting 'Daddy! Daddy!'

They're all there. Tom's detached from the others, as is his way: casting around, turning over stones, pulling himself up into trees, mooching. Every decent police officer, seeing how he moves, would have him in cuffs and back to the station on suspicion of behaving suspiciously. Now he's looking down at the hawthorn tree, and wondering why there are so many sparrows there all of a sudden.

My dad, after a wistful, benevolent look at them all, finds them too loud for his rather deaf ears, and has limped carefully down the hill to examine a water trough.

There's still plenty of cider left, despite X's help, and it'll take a while for this heat to go. 'Let's stay here tonight,' I suggest. Well, why not, they agree, on the promise of a fire, and sleeping bags to be delivered later.

So that's what we do: kidneys on sticks, spitting out the tubes; the cows with headlamp eyes as close to the fire as the fence will let them; the kids' latest fratricidal schemes laid aside because they can't get at one another across the flames; a fox come to watch; Jonny sick in the hedge; Rachel all mystical, snatching screaming leaves out of the fire; Jamie mesmerised and inscrutable; Tom at the centre of things for once, charring sticks for stabbing.

When the fire hunches down, they look up, and the more frightened they get, the more they start naming. They don't know many of the real names of the constellations, so they make them up: frogs, trees, fish, birds, hedgehogs, flowers, deer. All living things: there are apparently no shoes or tractors in the night sky. And they are all connected by story. The birds catch the fish; the deer eats the flower; the hedgehog dances with the frog; and then I wreck it all in the name of education by saying that over there, just above the treetop, is the Milky Way, which is home for us, and it has 500 billion stars (I write out the number in charcoal on a log: 500,000,000,000), and I think that that misty blur there is Andromeda, the only galaxy other than our own that we can see with the naked eye, and that it has a trillion stars, and it is 2 million light years away, so that light *there*, Rachel, is the oldest light you can see: it started its journey to this farm before your ancestor, *Homo ergaster*, first started prowling through the African bush. Then I get into my stride and start waxing poetic about whole solar systems being eaten by ravenous black holes, and electrons jumping like fleas between energy levels, and gas filaments millions of miles long, trailing through the heavens like the ripped veils of jilted brides, and the hearts of moribund stars continuing to beat and push light into the dark after the body has been crushed, and energy booming silently and billowing between galaxies. But they've all switched off and gone back to flicking marshmallows and burning holes in their wellies.

'Fifteen billion years ago there was nothing', writes the physicist Chet Raymo. 'Then God laughed. An infinitely dense and infinitely hot seed of energy sprang into being from nothingness and flowed instantly into matter.

According to current cosmological thought, that first laugh took a billionth of a billionth of a billionth of a second, and when it was all over, the universe was off and running.'

It's not a bad myth, but I've heard better, and I've certainly heard more useful. Presumably X and his son (who has now arrived) and my dad (who is still sitting on the log) can now give the definitive version.

I imagine that the Neolithic named the stars. It is unkind to blame them. If the Judaeo-Christian story is right, they'd been told to name all the animals, and for all they knew – and for all we know, come to that – the constellations *are* animals. Nor do I blame my children for trying to name, in their fear. Names are better than numbers. Names at least connote relationship. We can have a relationship with frogs and stars and hedgehogs and spiral nebulae. The number I wrote on the log means nothing.

Tomorrow it's midsummer. It was a crucial time for Neolithic people, and we'll mark it, as expected by many of my friends, by going up to some standing stones, drinking too much blackberry wine in leather cups we bought on eBay, playing some medieval tunes on our whistles and fiddles and mandolins, and trying to feel bedded in the tilting earth. It always has precisely the opposite effect. I feel closer to Simon Cowell than to the original occupants of the skulls in the long barrows, and I can't wait to get home and back onto YouTube.

This is worth knowing. For I've started to think that the Neolithic is comprehensible. Not only is it not comprehensible on the grounds that nothing is; it is not comprehensible because, although many modern modes of thought were forged then, the last 10,000 years or so have done terrible things to me. Back then they felt the world swing under their feet. I don't. They could rejoice that this was the lightest day of the year; I remind myself that it means that from now on the light will retreat.

★

Well, we went to the stones and stood around in the rain with lots of people in tie-dye, and I felt nothing and nothing happened, and so I went home and resolved to have a week eating just porridge and flatbread. So that's what I did, and it reminded me that most Neolithic mealtimes must have been very boring indeed. This sort of culinary boredom gets to you. I've spent long periods with Bedouin in Sinai and the Egyptian White Desert: their diet is more or less Neolithic – flatbread baked on an upturned pot. It drove me mad. But at least they have marmalade and tinned tuna.

In the course of our life as a species, we have eaten around 80,000 different animal and plant species. Yet just three species – wheat, rice and maize – are today the staples of three-quarters of the world's population. Did we choose to be dreary, or was dreariness forced upon us? Both. Sometimes we were under duress, and sometimes not. Certainly as hierarchies burgeoned and consolidated – founded first on kinship, with the unchallengeable imprimatur of the ruling ancestors, and later on political rather than biological dynasty – the regimes discouraged, often bloodily, the whole idea of choice. A citizen who opted enthusiastically for figs over wheat might decide to opt for a new ruler over the incumbent. 'How can you govern a country which has two hundred and forty six varieties of cheese?' asked Charles de Gaulle. If the passion for cheese is strong enough, you can't, and so all modern totalitarians – whether neo-liberal or communist – hate small producers and love monopolies. This can be seen from the late Neolithic onwards. Today just four companies control more than three-quarters of the world's grain trade, which is how our rulers like it. Even if the politicians aren't being paid off by the monopolies, it's easier to do business with the few rather than the many – particularly if you've been at school with the CEOs of the few.

The increase in human population in the Neolithic culminated in the critical masses for state formation. As population boomed, so did abstract thought and hierarchy. The reasons for the coincidence of these trends are complex, but the trends didn't bode well for individual humans in the rank and file of Neolithic society.

Population and hierarchy are causally related, and so are abstraction and hierarchy. Accelerating abstraction is clear in Neolithic art. The early Neolithic art of Iberia and Brittany is comprehensible: a bow is a bow is a bow. But as the Neolithic progresses, the art becomes far more difficult to interpret. It becomes, in other words, the province of experts, privy to the secrets that unlock the meaning. Them-ness and us-ness is there in the art. Sometimes the art was difficult specifically in order to keep it from the hoi polloi, but no doubt there were some genuine metaphysical advances that demanded more sophisticated exposition – and hence interpretation. Put this alongside the growing population – most of whom could not be priests or rulers – add the fear that comes from knowing that if that year's crop fails your children will die, and you have all the conditions for a toxic dictatorship.

In these conditions you'd expect increased violence too – both between members of one community and between communities. If your crop failed, but the village-next-door's didn't, the natural thing was to pick up your spears and your sacks and march next door.

That, says the consensus, is what happened. There is one high-profile dissenter, Steven Pinker, who is committed to the idea that things have been getting incrementally better (an idea we'll meet again in the final part of this book) and, accordingly, that the Neolithic was less violent than the Upper Palaeolithic. The Fall of Man, for him, was a fall *up*.

Before static societies emerged, he says, hunter–gatherers killed 15 per cent of the population, but when we settled we became less homicidal. The argument rests on very slender archaeological foundations. It's true that Upper Palaeolithic hunter–gatherers have a high incidence of bony injury, but most of these injuries can't confidently be attributed to humans. There are many other more likely candidates, such as falls and aurochs horns. To counter the orthodoxy that hunter–gatherers are relatively pacific, he cites examples of violent hunter–gatherers. Extrapolating from modern hunter–gatherers to prehistory is dubious, and he also relies on examples of hunter–gatherers whose whole sociology, economy and psychology have been disrupted by *states*.

Pinker also notes that there is dreadful intraspecific violence in chimps and bonobos. The further we get from them, he apparently believes, the more peaceful you'd expect us to be. It's a strange argument. When we became behaviourally modern in the Upper Palaeolithic, we were, in terms of our basic physiological and psychological wiring, as far from the chimps as we are now. What changed over the succeeding millennia were the circumstances in which our residual tendency to be violent was realised. In the Upper Palaeolithic – as in most subsequent hunter-gatherer communities – there were far fewer reasons (because there were far fewer people) to engage in the costly and dangerous business of trying to kill fellow humans. There was plenty of land to go round, and there were plenty of fish in the sea and caribou on the tundra. If someone annoys you, the easiest thing to do is simply to move on. Then, unlike later, it was easy to move on.

The Neolithic's increasing abstraction is part of this equation too. It's harder to kill a real person than the idea of a person. All murderous dictators know this: turn Moshe Cohen and his wife, Hannah, into generic Jews or, better still, 'parasites', or transmute their individual faces into cartoon Semitic noses, and it'll be far easier to get the mob to torch their house. If the abstract 'benefit-grabbing asylum seeker' is shown to be, in fact, Abdul Mohammed, loving father and devoted son, who likes playing chess and the ukulele, it'll be far harder to get him trussed up and dragged back onto the plane to Syria.

Population plays into this: individuals are harder to see if there are lots of them. And the same effects are at work in the exponentially increasing commodification of humans. As profits became more important, individuals became valued increasingly not for who they were, but for what they could do. And if they couldn't do it, well … It's the old, old and new, new story.

As population increases, society becomes more complex and there is a (perceived) need for regulation. A chasm opens up between the regulators and the regulated. If the market does all the regulation, the rich are at the top and the poor at the bottom.

The most useful information about ancient violence that contemporary anthropology can provide is probably about the effect of alienation from the wild. We are all wild creatures, even if we live in Manhattan and eat only microwaved meals from a plastic pot, but we have become to some limited degree habituated to our modern life behind bars. For the first few generations in the urban zoo, though, it is psychologically very tough. Mental illness, suicide and violence are rife among hunter–gatherers who have recently stopped hunting and gathering. '[T]he human spirit has a primal allegiance to wildness,' says the writer Jay Griffiths. 'The first commandment is to live in fealty to the feral angel', and when that commandment is breached, retribution is terrible and immediate. It involves, first, exile to a game-free, fish-free, desertified 'wasteland of the mind', and from there the former hunters set out to hunt themselves and other humans. In Africa, Australia and the Arctic indigenous people told Griffiths that the only solution to the problem of violence and other anti-social behaviour was 'the land' – by which they meant the very opposite of what nationalists mean: they meant the natural entitlement of all wild people (and that's *all* people) to all the wild earth. The Yolngu people in Australia call the land 'mind medicine': an Inuit musician called Jimmy Echo told Griffiths that 'violence comes from being outside nature'.

I don't need books or articles to tell me this. Our family life is one long proof. Take the children into a green place, and immediately the knives are sheathed and the meannesses abate. It's not just a question of distraction, or burning off energy. Take the same children, in the same state of fratricidal excitement, to an indoor entertainment, and the civil war continues. You'd think they might be inhibited by the trappings of urban civilisation – uniformed museum staff, for instance, or the sternly moustachioed waiter in our local Greek taverna – but not a bit of it. Trees make moral demands on them that no policeman ever could.

There was one key corrupting element that the Neolithic did not acquire – the weapon of mass spiritual destruction that is written language. Written language consolidates the hold of dynasties: it writes succession, debts and formal obligations (sometimes literally) in stone, allowing them to prevail against justice, mercy and natural evolution. It is the apotheosis of abstraction: the decree nisi of the divorce of humans from the natural, concrete world (the Enlightenment, as we will see, was the decree absolute). To write something down gives it a wholly spurious authority – an authority that trumps the authority of experience. If your world is there in characters on a clay tablet, there's no need to go outside to the woods to verify anything; the woods won't be given the chance of contradicting what's there. And – how glorious! – the person who wrote those characters, who created the whole world, is *you*!

The earliest writing was Sumerian. It was pictographic, and still nodded deferentially to the world outside human heads, but soon the pictures were ousted by the *lines* (yes, lines again) of Sumerian cuneiform – the aggressive, rapacious stabs of a man-made, man-held stylus into natural clay, with which Mesopotamia remade the world in its own image. Purely oral cultures had demanded relationship – between the teller and the hearer and the teller and the source. To gather a story, an oral storyteller has to go out into the wilds of experience, tune their ear to the tones in which the wild, if it is in the mood, chooses to render the story, ask the source's permission to pluck the story, pledge not to use it in a way contrary to the source's intention, bow as they walk backwards out of the wild, as the Hasidim walk backwards from the Western Wall in Jerusalem so as not to slight the presence that sits there, tell the story faithfully around fires and bind listeners, on the pain of societal death, to keep telling the story.

When there's writing, it's different: someone sits in a room making eternal, binding lines from the thoughts in their head – lines that derive from no authority other than the head, lines constraining the future action and orientation of others just as surely as the line of a fence stops sheep from grazing where they want. To inscribe a list of debts or to detail a treaty is to tell a story just as much as to relate how that rock is the claw of a giant toad, or how your father gathered talismanic leaves, smoked a foul pipe and carried on leaving a vapour trail of coal tar soap long after his body was burned.

There was another important and catastrophic stage in the creeping hegemony of written language. It did not happen in the Neolithic, but its seeds were sown there. It was the advent of alphabetic – phonetic – writing. Pictographs nodded to the non-human world, relying on a sketch of a tree or an ox to convey meaning. Phonetic writing severed language's connection with and dependence on the natural world, and humans for the first time began to believe that language was a uniquely human possession. Until then – and throughout the Neolithic, despite all the charges I have levelled against it – the non-human world spoke and listened, though its accents were increasingly hard for humans to hear, and its stories increasingly disregarded, patronised and supplanted. Not until the alphabet was it presumed that the natural world was dumb. The pre-alphabetic natural world had sung, versified and pronounced. With the alphabet it fell silent. And you can do anything you want to dumb beasts, can't you?

★

We're in Wales again, with Burt and Meg. Meg says she's forgiven me for my springtime sermon about mastery, but I'm not so sure. But it's all very jolly here, as it always is with them, and we've been herding sheep, feeding donkeys, fiddling with the hydroelectric turbine, swimming in their river, looking for the bones of Iron Age priests, doing an anatomical jigsaw puzzle with a turtle's skeleton, mapping badger latrines and boiling up some Herb Bennet for children's diarrhoea and some coltsfoot for an ulcer on a neighbour's leg, and now we're back at the house eating lamb chops and spinach. ('That spinach must have the highest iron content of all time,' says Meg. 'It's fertilised with my menstrual blood.')

Burt looks out of the window, slams down his plate, flings open the door and bolts outside and down the lane towards the valley. We race after him.

'It's started,' he shouts. 'Quick!'

The wind is blowing leaves from an oak tree, and Burt is

trying to catch them before they reach the ground. It's very good luck if you can, he says.

Jonny refuses to join in the game. 'The leaves are meant to fall,' he says solemnly. 'This is their time for being on the ground.'

'*La li-li-li, li-li.*'

AUTUMN

'The land of a rich man produced abundantly. And he thought to himself, "What should I do, for I have no place to store my crops?" Then he said, "I will do this: I will pull down my barns and build larger ones, and there I will store all my grain and my goods. And I will say to my soul, Soul, you have ample goods laid up for many years; relax, eat, drink, be merry." But God said to him, "You fool!"'

Luke 12: 16–20

'This year,' said my lovely friend Liz, who makes driftwood sculptures, felt slippers and fairy tales on a smallholding that looks to the Outer Hebrides, 'we're not going to go gently into that good night. We're going to make a hell of a lot of noise and drink a hell of a lot of wine and eat a hell of a lot of what the summer has given us, and head into the autumn grateful and sore-headed.'

So that's what we do. We all pile into the car and drive north, breaking the journey in a motorway hotel near Glasgow which has special rates for adulterous flooring tile reps, and a shoe cleaning machine like a spinning hedgehog, and by lunchtime the following day we drop down the little road into Liz's glen. We have the sore heads already, courtesy of the hotel's bulk tank Shiraz, but the rest is as promised: a sheep turning over the fire, salad from the hedgerow, dancing to the fiddle, the accordion and the bodhrán, recordings of Indian ragas and whale song to fill the gaps between tunes, vats of rhubarb wine, obscene and sublime songs, swimming in the sea to cool down ready for the next reel, a sulky crow sitting on a tree and complaining about our behaviour, and a rumour of dolphins.

It's wonderful. But, dear Liz, I have a problem. It's not meant to be like this. Dark bleeds into the light; the light doesn't suddenly shut down. The whole idea of the equinox and the solstice depends on measurement, calculation and a division that isn't really there. Berries don't spring out overnight in the autumn to replace the heavy heads of grain brought down by your sickle. We shouldn't go to megaliths expecting things to happen: we should expect things to happen all the time. The dark, even in the north of Scotland, is only ever relatively un-light, and in the deep dark, fire, sent originally from the sky in a lightning bolt, keeps the day burning. Since the fire came from the heavens, we might say that the summer sky never shuts down: it just relocates to the hearth.

Your beautiful green cake representing the summer, and covered with marzipan flowers, is sliced up and ceremonially eaten. I understand why, and you are brave, good and generous, but to destroy something beautiful and expensive, whether by eating or breaking, reminds me uncomfortably of the Neolithic destruction of pottery (embodied in modern Greek plate-smashing) to show wealth and status.

In the hunter–gatherer world the dark inched in; no day was much shorter or longer than its neighbours; each season gave enough, in many different ways, and it was the daily *enough* that was daily celebrated, not the filling of barns.

In the modern West we have barn-filling celebrations: harvest festivals. We love them. It's good for us to be grateful, even if we can't coherently say to whom the gratitude is due. Harvest festivals seem appropriate. I wonder.

'We plough the fields and scatter the good seed on the land,' sing the children at squirrelling time. (Well, no they didn't, but let's let that point go.) The hymn is rather self-congratulatory. *We* made the right agricultural choices, *we* put in the graft and so *we're* going to have a full belly this winter, and a great knees-up in the village hall next week.

There's a show of humility in the next line: 'But it is fed and watered,' we inform God, 'By thine almighty hand'. It doesn't wash. God is simply a partner in the project that *we* have conceived and, in the traditional economy, he and his priests might be entitled to a small cut – say 10 per cent, the tithe – for his pains. We're to be clapped on the back, too, for backing the right God, and for having appeased him in a way that's evidently worked. It's very Neolithic, and very un-Palaeolithic. Upper Palaeolithic thanks are constant, trembling and directed towards the animal or plant that's given itself up. And they can't, unlike Neolithic thanks, be redirected sycophantically to the earthly ruler who is God's deputy, who has directed that the seed be sown, on whose beneficence you depend and who is sitting in the padded pew in his best suit, ready to receive with an avuncular smile the grateful tugging of your forelock as you file out of church.

★

'Now we can all die happy in the cold,' beamed Liz, as we hugged goodbye.

The earth does seem to die when the sun retreats, and in past winters I have put my hand and my ear on the ground, trying to convince myself that there's a residual pulse, and felt and heard none, and felt the dark grow in me.

That's a Neolithic thought: no sun, no life; no crops above ground, no life – hence the desperate focus on the winter solstice, when the sun will once again start to be on *our* side. Walls and fences – a binary world, unimaginable to hunter–gatherers, but inhabited by most of us ever since: sun/no-sun, light/dark, dead/alive, on/off, God/no-God, believer/unbeliever, black/white, us/them, clean/unclean, feast/famine. Yet the Neolithic wasn't as far gone as we are. It knew, at some level, that death was creative; that next year's harvest was being knitted together in the cold dark, that the winter was alive. I've forgotten even that.

★

Two sets of plates are doing the rounds in front of the big house. On one there are little canapés, topped with smoked salmon and dusted with caviar. On the other there are sausage rolls from the supermarket down the road. The canapés are offered up deferentially to the red- and black-coated riders, the sausage rolls pushed grumpily at the foot and car followers and the crumpled-berry-faced men in green who will follow the hunt on quad bikes with terriers in a box on the back. The riders get cherry brandy in wine glasses (cut-glass for the Master of Foxhounds), the pedestrians cooking sherry in paper cups.

This is one of the most popular meets of the year. From the terrace of the Palladian house you can see thousands of acres of the sort of country that makes the expatriate English dewy-eyed: a verdant patchwork quilt of fields grazed by cows that are building the roast beef of Olde England; trimmed blackthorn hedges; remnants of woodland in hilltop redoubts; everything rolling, gentle, understated. All stolen from the peasants in the eighteenth century, of course, flooding the cities with cheap, desperate labour. But let's not spoil this glorious sunny day with thoughts like that. And glorious it certainly is, from the dappled flanks of the pacing hounds to the jodhpured thighs of the stockbrokers' wives.

The Master, everyone agrees, is on splendid form this morning. She cut her teeth in a tax lawyer's office in the Cayman Islands, and then used them to savage business rivals and over-zealous revenue inspectors before sweeping back in glory to London and the shires to take possession of a surprised widower (who had no chance once she had him in her sights), a new-build country house of dubious taste, a document in her safe giving her immunity from prosecution and thirty couple of foxhounds with a history going back a couple of hundred years. Now she sits back in her saddle as in her boardroom chair, glass in hand, thanking her hosts for their hospitality, making some laddish jokes *just* on the right side of decency, and wishing everyone (apart from the fox, ho ho!) a great day's hunting.

The little hard huntsman has been listening dutifully, his eyes flicking between his hounds. He knows them as well as he knows his children. Is Darter going lame on that left hind? Did Chanter get too much of that old cow in kennels last night? Don't want her to be slow today. We'll need her nose with scent like this: it's too dry for comfort. He is too professional to give any sign of what he thinks of the Master. His father was huntsman before him. He could ride before he could walk. He dreams of foxes and has a red-haired foxy wife who makes soup from the hunt's dead horses if they're fresh enough. If you ask him if he likes foxes he smiles, fingers his chin, and says, 'Well, that's complicated, that is', as indeed it is.

Today he's going first to Brandy Wood, which we can see from the terrace. There are plenty of foxes there, dug in between the beech roots. It was here that the huntsman's father, cheering his hounds onto a big grey dog fox, rode into a low branch and snapped his neck. The huntsman tries not to take it personally, but when he's in Brandy Wood some say that his jaw is set harder, and that he tries harder to kill.

Tom and I are there, on foot, because we spent the night with a friend not far away, and I thought he should see what all the hunting fuss was about. We're planning to make our way after the hounds as best we can, and then perhaps hitch a lift back in a horsebox.

The huntsman raises his cap to the Master and the hosts and blows his horn. The hounds' heads come up; they thrill and shiver at the thought of work, and trot off down the drive after the huntsman. There's a general cap-doffing and girth-tightening and rein-gathering, and then the cavalry clatter off too.

And it's then, next to one of the trestle tables, reaching out behind one of the last horses for a sausage roll, that I see a very strange arm. It's covered in the long, coarse hair of a red deer. The hand at the end of it is nearly black, with closely bitten nails and the power of a hydraulic vice. Yet it picks up that sausage roll with the delicacy of a feeding moth.

The horse moves away now, and I see the face that's being filled with sausage rolls – for there are many in there already. Like the hand, it is very dark. But little is visible behind a black beard, flecked with grey, cut into uneven steps, and under a round cap of what looks like otter fur, pulled down to his eyebrows and ringed with black crow feathers. His nose is straight, long and dripping. As I watch, he crams in another sausage roll, presses one nostril with a thumb, and blows hard down the other, spraying the table. There are some canapés left: he picks one up, sniffs it and grinds it underfoot. He's in a deerskin parka that falls below his waist.

On his legs are thin breeches, perhaps made of the shaved, brain-cured skin of a stillborn animal, and he has badger-pelt boots, with the face stripes running down towards his toes. If he laughed, which I think he often does, it would be a noise of crushing solemnity.

He is shorter than I am by a good head but has a power about him that gives him the apparent height and breadth of an old oak. It's a centripetal power which pulls everything – eyes, tables, ideas and sausage rolls – towards him. He's like a black hole with a beard. And that is why, for a while, I don't notice another figure by the table, clad too in skins, but younger, paler and rather wan. This boy is uninterested in the food: he's transfixed by something behind me. I manage to slam my eyes shut, drag my head away from his father (I'm sure that's what the bearded man is) and turn round to see what so fascinates the son. And there, just behind me, is Tom, looking at the table as if he wants to bore a hole through it with his eyes.

'Tom?' I say. 'Tom. What are you looking at?'

And Tom shakes his head like a dog coming out of a pond, looks quickly at me, and says, 'Nothing.'

As you please.

From the fields below the house a hound is whimpering, and I can hear the huntsman shouting encouragement: 'G'yon, Chanter, find him, find him.' And the other hounds, knowing Chanter's nose too, are rallying round. 'Lieu in there,' urges the huntsman. 'Lieu in.' And then suddenly, in a crash of sound like a breaking wave, they are streaming straight up the hill. The huntsman's misgivings about the conditions were misplaced. They're screaming along on a breast-high scent, Chanter falling back now as younger hounds take it on. Surely this isn't an old drag left in the early morning by a fox returning home after a night's foraging. Surely this fox has been caught out, sleeping under a hedge.

And yes! There he is, a field away from the lead hound, running strongly with his brush straight out behind him, confident in his speed, not bothering to be smart, making straight for the old fox fort in Brandy Wood.

'Whoo-whoop!' shrieks the huntsman. 'Whoo-whoop!' And the hounds, who know what it means, double their pace and volume. Behind the huntsman the Field Master is trying to hold back the thrusters. 'Give hounds room, please,' he yells. But it's no good. Horse and human nostrils flare, hats are jammed on hard, hip flasks are whipped out for a last fortifying gulp, expensive and quivering limbs prepare for take-off, spurs prong bellies and crops thwack hindquarters, and they're off, mud flying around their velvet coats.

The oak man, mouth full, spins to face the spectacle. His arms drop to his sides. He leans forward to peer and pushes back his cap, and his eyebrows twist in furious disbelief. He punches the boy on the shoulder, motioning with clenched fists to the field.

The fox is still well ahead, but the gap is narrowing. His brush is starting to droop; his legs are heavy with earth.

The huntsman is a field behind the leading hounds. A whipper-in is urging stragglers on, 'Hark forrard, hark forrrrrrrrrd', and a high hedge has sent several followers over their horses' necks into the good soil of the English Midlands. There will be fewer intact collarbones around the dinner tables of Tory Leicestershire tonight. The quad bikes are buzzing along a lane towards the wood, their boxes yapping.

The fox tries to push through a hedge. He nearly makes it, but it's just too tight. He runs along the ditch at the bottom of the hedge until he reaches a fence. He jumps at it, scrabbles at the top, then falls back. He's tired. He is losing his lead. He has another go at the fence, and this time he's over. Behind him, Darter is too excited to bay: and there's no need now: everyone can see the fox. It won't be long now. The terrier men won't need to unbox their dogs or unstrap their shovels. The hounds will roll him over in the open, as the old songs say.

But wait: it's not over yet. That lead hound is puzzled. So are the others who join him. Even Chanter, now labouring herself, can't work it out. The huntsman comes up. He scratches his head. Can the fox have crawled into a drain? A big rabbit hole? Surely not. He swings off his horse, holding it by the reins, to look into the ditch. There's nothing to see. He remounts, toots his horn and casts his hounds in ever-widening circles around the point he last saw the fox. Nothing. Very odd. Perhaps some idiot spread slurry there that morning, drowning the scent, but it doesn't seem like it. To have a run like that, and then lose him right at the last minute. It was typical of Brandy Wood. It never plays by the rules. He hates the place and everything in it.

And then, just as he's about to give up on this fox and try for a new one, there's a screech from one of the quad bikes. 'There he goes, John. By the barn.' And so he does. A good quarter of a mile off. How he got there no one could fathom, but it won't save him. He's slinking slowly along, further away than ever from the wood, his white belly-fur dragging along the ground, his head down and his tongue out.

'C'mon my lovelies,' says the huntsman. 'C'mon. Let's finish the job.' And standing in his stirrups, he beckons to his whippers-in to keep the followers back, and lays the hounds on by the barn. Away they go once more, the fox heading for the beech-tree earth that the earth-stoppers can never find, the shovels can never violate, and which is shared by badgers who'd make the terriers think twice. And the hounds: well, the hounds go for the scent and the pride and the camaraderie and to satisfy the ancient dream, from which they can never truly wake, that says that they are really hunting a reindeer.

The fox will never get to the beech tree in time. He's going to be killed by brindled half-wolves who think he's something else.

The oak man unfreezes. His flesh, of which there is a lot, seems incidental to his presence. He grabs his son by the arm and suddenly they are running together towards the wood. As he unfreezes, he seems to freeze the action ahead. Then the hounds (my memory insists) are pulled back towards him, as were my eyes and my thoughts and the sausage rolls. Only the pair are moving forward, running with an easy lope. They're past the huntsman, they're past the lead hound, they're up to the fox. And the oak man leans down and scoops up the fox and runs on with him cradled in his arms.

At the edge of the wood he and his son turn and look back towards us. The oak man raises a hand. I hope it is a salute. The son does too, and out of the corner of my eye, which is where you see everything that's worth seeing, I see that Tom's arm is raised.

The pair and the panting fox go together into the wood. The huntsman gathers his hounds. He blows his horn, but it's not that we're hearing. There's another sound: '*La li-li-li, li-li.*'

I half-expect a melodramatic ending: perhaps the huntsman, led on by the strongest-scented fox in the annals of hunting, will die on the same branch that killed his father. It doesn't happen. It's an ordinary, unspectacular no-kill day after a much-appreciated reception on the Palladian terrace. Foxes often go missing. Nothing odd about that.

The Master goes off for a charity dinner with some strategically important friends, a number of the mounted followers go to hospital, the men on the quad bikes go to the pub to work off their frustration with darts and lager, and the huntsman, after his bath, sits by his fire in his house by the kennels, staring at the sputtering logs with that smile he has if he's asked about liking foxes.

★

We're back in Derbyshire, at Sarah's farm, where we first met X and his son, where Tom got into the habit of offering food for food at a woodland altar, where there were solid columns of coal tar scent between some of the blackthorns; where I starved and saw the shimmering; where the magpie tick-tocked and the hare lay shameless in the moonlight; from where we set off to hunt caribou that had died 40,000 years ago.

The rest of the family have gone off shopping in Bakewell, leaving me and Tom sitting at our old camp in the wood, looking across the dale.

Tom's talkative today. 'If I farmed that' – and he sweeps his arm to take in the whole of the hill, 'I'd put sheep up the top – it's wasted at the moment. And I'd fence off the pool – it's full of lead – and I'd pipe water to troughs just below those trees.'

Would you, Tom?

I wonder what you'd do if senses other than your sight affected your decisions. I wonder if the smell of the hill would change your mind, or whether you'd find, if you slept up there, that the *hill* had a view about sheep or troughs. I wonder what would happen if you let the hill plan you, rather than you planning the hill.

'Would you like to farm here?' I ask.

'It'd be fun. It'd be hard, but fun. And if we'd been here as a family for ever, I'd feel that I *should*. I'd want to, as well. Wouldn't you?'

I don't reply because there is too much to say.

I look at him. He has astonishing poise – far, far more than I had at his age. He knows who he is. He's about to enter the biting winter season of adulthood. Something has to be done.

Before it's too late this poise needs to be shaken and replaced with something that will last. 'When I work with troubled youth, I try to make them more troubled,' says Martin Shaw. That's far more important with untroubled youth.

★

'Tell me a story,' says Tom.

'I've got two,' say I. 'The first is in my pocket.' I pull out a crumpled news item from one of the daily summaries sent to practising lawyers.

> According to *The Lawyer*, the UK's largest law firms are squeezing their staff and fee-earners ever more densely into offices, with the space allocated per person having shrunk by more than a third in recent years. The average amount of space per square foot for each staff member has fallen 33%, for fee earners by 32%, and for partners by 9.6%.

'Fascinating,' says Tom. 'Hope the other one's better.' Not really, to be honest.

> *Once upon a time there lived a very rich man and woman. They had been given (the story is not clear by whom) a huge estate to live in. It stretched in every direction as far as the eye could see. It was so big that they had never been to the edge of it, and nor had anyone they knew. It was*

very beautiful. Wherever the man and the woman were, the views were amazing. There were woods, mountains, rivers, lakes and valleys, and it was full of friendly animals. The rivers teemed with fish, the trees chattered with birds (and indeed the trees themselves chattered, but much more slowly), and there were great herds of deer and wild cattle and wild pigs. Although there were so many animals, the man and the woman knew the names of all of them, and if they called a deer by name, the deer would come and nuzzle their hands.

There was plenty of food throughout the year: berries, flowers, mushrooms and even, though they didn't taste as interesting as the other things, the heads of some big grass stalks that grew on the edge of the wood. And of course there were animals. If a deer or a fish knew that its time had come it would say to the man and the woman, 'Eat me and enjoy me,' and the man and the woman, tearfully, would thank the animal and accept the gift.

The man and the woman were fit and healthy, for the climate and food were good, and wandering across the land kept them strong and supple. They moved from house to house, leaving the wind and the mice to clean up after them, and so every night they had a fresh lodging.

But one day (no one knows quite how it happened) the man said to the woman: 'I'm fed up of going out every day and having to collect food. And why should we have to wait until an animal decides it's going to die before we have meat? Let's cut down some of the forest, build a wall around the part we've felled, build a house inside the walls, plant some of the big-headed grass inside the walls, and bring some of the animals inside too, to kill whenever we want.'

The woman thought that this was a very bad idea indeed, and said so, but the man wouldn't change his mind. So that's what they did.

As they took their stone axes to the trees, the trees groaned, the birds were outraged and the deer came to ask what was going on. But though the woman wept and heard the groans of the trees in her sleep, the man was unmoved. When he was lifting stones for the wall he wrenched his back, which made him even more bad-tempered than ever. The wall was high. It meant that from their house (which quickly got very dirty) they couldn't see the mountains or the trees: all they could see was the wall.

The animals didn't like the idea of living behind the wall, and wouldn't come quietly. So the man snared them and dragged them in, and they screamed so loudly that even the man heard them in his dreams. They didn't give these animals names: they were just 'it'. Because the animals couldn't graze where they wanted, the couple had to grow and gather fodder for them, and clear away their dung. The couple themselves got bored of the grass heads. It was hard work grinding them, and the woman got a bad back herself from doing it, as well as a nasty fever that she caught from the deer and couldn't shake off.

The man and the woman had lots of children, and their children had lots of children of their own.

After a few years of this the man said to his wife: 'I wonder if we made a mistake? Shall we go back to the way we were?' The woman looked sad. She knew that her children and grand-children had killed most of the animals in the forest outside the walls, and that they didn't know one mushroom from the other, and that the remaining animals now distrusted them and wouldn't come to them to die.

'We can't,' she said. 'Though I'd love to. But at least we can try to say sorry for what we've done.'

So they went out into the blackened, scrubby bits of forest, looking for something that would receive their apology. They couldn't find anything at all.

Neolithic: Autumn

And that, they decided, was a problem even greater than the grass-head diet, the deer-fever, the knee-deep dung, the bad backs and all the mouths that had to be fed.

ENLIGHTENMENT

'When a well-packaged web of lies has been sold gradually
to the masses over generations, the truth will seem utterly
preposterous and its speaker a raving lunatic.'
Dresden James

'The general materialistic framework of the sciences at
the moment is not wrong. It is simply half right. We know
that mind is mattered. [A huge number of stories] suggest
[…] that matter is also minded, that this mindedness
is fundamental to the cosmos, not some tangential,
accidental, or recent emergent property of matter.'
Jeffrey J. Kripal, The Flip: Who You Really Are and Why It Matters

'"There is a mulla-mullung. He is white and he is
mad. I look into him and see no Ancestor, no Mami-
ngata; nothing is his flesh, he has no Dreaming."
"No Dreaming!"'
Alan Garner, Strandloper

'It's all in Plato, all in Plato: bless me, what
do they teach them at these schools.'
C. S. Lewis, The Last Battle

I used to think I knew what animals were. They ate, moved, and often if you looked at them they looked at you. They had purposes of their own, and seemed to be far more disciplined and energetic in pursuing their purposes than most humans I knew. Many of them had very desirable, exciting addresses: treetops, cliffs, caves, the open sea or the air beyond my sight.

It was quite easy (I thought) to give a basic description of a dog, though impossible to give a full account of either dogs in general or any particular dog – just as it was impossible to give a full account of anything interesting or important. Dogs began with their noses and ended with their tail tips. They were covered in fur, had four legs, two eyes, liked chasing cats and rabbits, and ate pretty much everything – though they were particularly fond of meat. They could be fierce or affectionate.

One day, though, I discovered that there was no such thing as a dog – or indeed any other animal. That was my first working day as a veterinary student.

All the new students put on their new white coats and filed nervously into the dissecting room. There was an overpowering smell of formaldehyde. Soon we didn't notice it: it was our new air.

There were many metal tables. On each of them was what I used to think was a dog: a greyhound, in fact. Six of us were allocated to each 'dog'. These dogs were here because they couldn't run fast enough. They were ideal for dissection because, though they were too slow to be economically worthwhile, they were lean and muscular. Very little fat obscured their muscles. And for that first term we did muscles.

We started that day, and continued for a while, with the

shoulder. We learned the Latin names of the muscles, their points of origin and insertion, and the mechanics of their action. I'm still great on shoulders. By the time we got to the hip I was tired.

It didn't matter, though, for shoulders and hips weren't connected in any way. They were separated by two large body cavities. A typical exam question would ask just about the shoulder, or just about the hip, and as long as you knew enough to answer five questions out of eight, it was fine not to know that dogs had hindquarters.

So we cut up the dog into shoulders and hips and tongues and lungs and brains and bladders until there was nothing left of the shape of the dog. With our scalpels we killed all dogs, live and dead.

It got even worse than that. Every fortnight, with a sense of grave foreboding, I'd climb reluctantly up an ancient staircase to a beautiful room overlooking the River Cam, sit on a sofa, and for an excruciating hour my ignorance of the chemical reactions inside cells would be cruelly exposed.

'I'd like you to write down, gentlemen,' – for we were all male, if not all gentlemen – 'the structural formula of the compound produced when the Krebs cycle moves two stages on from cis-Aconitate.' So I'd dredge though the sludge at the bottom of my mind and sketch something out on my pad, and the supervisor, often in his dinner jacket and black bow tie, would stalk round with his hands behind his back, peer at my dismal effort, snort contemptuously and say: 'No, sir! Try again, sir!' And so I did, painfully and pointlessly, until at last, out of a desire to get to dinner rather than out of mercy, the supervisor barked the answer and let me creep out into the night.

This supervisor made no secret of his contempt for the anatomists. 'I deal with *origins*. With *fundamentals*. A liver is just a factory in which *my* employees happen to work. Who cares if it has three lobes or three hundred? And' – he saw himself mainly as a molecular geneticist – 'knees are only the way they are because *my* genes say that they must be like that.' It wasn't just that there were no dogs: there weren't even any shoulders or hips. There were only the molecules making up genes.

It got no better when we started seeing live animals and learning how to treat their problems. Because no *animal* had any problem, ever. A shoulder or a kidney or a biochemical pathway or a chromosome or a gene might have a problem. We never treated animals: we treated problems and pathways and genes. Animals didn't exist.

Near the beginning of this book I admitted that I'd never seen a tree. Now, near the end, I admit that I've never – at least as an adult – seen a dog. It's a shame.

*

I put the noose round my neck, knot it and pull it tight.

X and his son, sitting on the bed, look alarmed. X gets up to stop me, then seems to remember that that's not within the rules and sits down again. They've been more or less constantly in sight since I got back to Oxford, and they're as close and solicitous as dead hunters can be.

I tug the collar down over the tie, put on a jacket, shout that I won't be late back and leap on my bike for the five-minute ride to dinner.

Tonight I'm dining in a grand old college with Professor Black. He had assured me that my project was futile: that no one could get close to the mind of prehistoric people. Now he's invited me for venison en croûte and old claret so that I can admit in front of his friends that he was right.

I'm late. I sling my bike against a wall topped by a martyred saint – one of the protectors of this place – and run through the cloisters to the Senior Combination Room. I've never discovered what they're combining there. Not different views, to be sure. Everyone seems to believe the same thing about everything, and to think that everything's very straightforward.

The Professor has assembled a small committee to appraise the exhibit he's dragged in from the Stone Age: a neat-bearded Shakespearean scholar with an international reputation and an Edinburgh accent designed for talking about the texture of scones and the soiling of doilies, and a little wizened physiologist in a crumpled suit who spends his life waltzing forlornly with the molecules of rat neurotransmitters. The Professor himself is a shiny, well-scrubbed Cockney sparrow of indeterminate specialty – perhaps politics or social policy or some dialect of economics: 'Don't try to box me in,' he told me when I first met him. 'The world's my oyster.'

It starts harmlessly enough.

'So,' says the Professor, handing me a glass of sherry. 'How was the mammoth and chips?'

The Shakespearean and the Physiologist had been well briefed, and lean in expectantly for my answer.

'Not a patch on what we're in for tonight,' I reply, keen to show that I can do banality as well as the next man.

'Well, that's progress,' laughs the Professor. 'You told me that cavemen lived in the best of all possible worlds.'

I'll not be drawn so easily. I know I have to sing for my supper, but I need to warm my voice up first.

'They had nothing on their walls like that,' I say, pointing to a late eighteenth-century landscape painting, showing tumbling waterfalls, leafy arbours and a recumbent swain.

'Lucky them,' says the Shakespearean. 'Romantic tosh. Give me Lascaux any day.' At which point the merciful gong goes for dinner, and we file upstairs, past more Romantic tosh, into the candlelit hall, where the undergraduates, hungry after their onerous days applying for hedge fund management jobs, stand waiting for us, black-gowned like crows with expensive haircuts.

A college scholar in a longer gown recites the Latin grace, thanking the thundering God of the Hebrews, his subsequent dreadlocked, anarcho-syndicalist incarnation and the college benefactors (many of them slave-traders) for the bounty about to be served up by the waiters. Then we sit down.

'You look distracted,' says the Shakespearean.

'I am,' I reply. 'I'm thinking about what you said about the picture.'

'It was a throwaway comment,' he says. 'Don't let it get in the way of this wine.'

Throwaway comment or not, it's not a comment that could be thrown away. It signifies a lot. 'What's the problem with the picture?' I ask.

'It's so boring,' sighs the Shakespearean. 'Nothing's happening. The only human in it is asleep.'

There we have it: The main attitude of humans towards the non-human world since the Neolithic, seen in both visual and written art. There have been exceptions, but nature has generally been seen as a stage on which human dramas (which are all that *really* matter) are acted out: as a backdrop to human affairs. Rarely does it matter in its own right. Sometimes it is the dwelling place of deities who have to be placated. Sometimes, through waves, bolts of lightning, earthquakes, prophetic birds or prophetic bird entrails, or gigantic prophet-swallowing fish, it is the means by which deities or disembodied Fate deals with humans; but where this is so, what matters is the end result for the human, or how it can be dodged.

Usually nature is simply ignored. There is little landscape-painting or birdwatching in Homer. The ancient Greek passion for understanding the natural world is driven by a love of order and system rather than a love of flowers and frogs – let alone the Upper Palaeolithic sense of one-ness with them.

Aristotle, of course, is an acute and diligent naturalist, but though it may be unfair to judge him by his lecture notes (we have nothing else), his concerns seem to be entirely anthropocentric: learn about the great outdoors, he seems to urge, because that's part of the philosophical curriculum, and you'll need the birds and the bees on your CV to be certified as a qualified eudaimoniac. We have him to thank, too, for the beginning of the *systematic* denigration of the non-human world, for he identified three types of soul, in a distinct hierarchy. Plants have a 'vegetable' soul; all animals have a vegetable soul and a 'sensitive' soul; and humans have vegetable, sensitive and 'rational' souls. I needn't explain their characteristics here: it's enough to say that this system embodied the hierarchy of being that is absent from most hunter–gatherer thinking, implicit in the Neolithic subjugation of animals and plants, codified in the Genesis injunction to subdue the earth (and unmitigated by Genesis's insistence that this subjugation is really a command to assume an often crushing burden of *stewardship*), and leads to Steve the Peedo and a Big Mac with fries.

Aristotle's attempt at a scientific zoology was, in any event, a flash in the pan. No one, so far as we know, builds on his work for another 1,500 or so years. From Herodotus onwards we have bestiaries of fanciful creatures, used as vehicles for moral points or merely amusing stories.

The Romans – rather surprisingly – seem rather more interested in the natural world for its own sake. Lucretius and Pliny the Elder would grace any modern natural history society, and just compare Virgil's lush landscape painting in the *Aeneid* with Homer's apparent blindness to anything other than the strategic importance of the land. But then we have to wait until the mid-seventeenth century for any real admiration of nature. The Renaissance and the Middle Ages reserved their praise for 'cultivated, productive land; for meadows, crops, gardens, orchards and ponds, not for nature in the wild. There was a disdain for untamed nature', which was associated in Western Christianity with 'brute beasts'.

Animal passions were the enemies of true piety. They had to be fought, and for many it followed that nature had to be fought too.

The English cleric and scientist Thomas Burnet, writing in 1681, summarised the consensus and set the agenda for the debate with the Romantics that would erupt in the eighteenth century. Natural landscape was dangerous and corrupt: its jaggedness showed God's displeasure with the fallen world, a world 'from which much of God's original perfection, the smoothness and symmetry, has been lost'. Good (Western) Christians, anyway, had better things to do than look at birds. Their real home was not here: their destiny lay in heaven, and so they'd better concentrate on getting there. Lines once more: a straight and narrow road leads to salvation; winding roads, of the sort created by rivers and badgers, lead to destruction. Tangledness is of the devil, for ensnaring the unwary. The devil himself is like a roaring lion, prowling around, seeking souls to devour; woods are full of wolves and sexual temptation; cosmic time moves in a straight line from creation, via Calvary, to apocalypse; and our human lives move from birth to death to (if we've stayed out of the woods) salvation. Body-hating Augustine, promulgating his doctrine of original sin, had taught that the natural process of birth brought sin into the world, and the Western church believed him. When the Reformation preached the supreme value of the individual's interior life – and the consequent inconsequence of the wild forests and moorlands – nature finally lost its battle for the heart of the West.

Romanticism protested its love of nature in flowery odes and sonnets, but the love was often onanism. The late eighteenth and early nineteenth centuries saw a brisk trade in coloured pocket lenses through which intrepid walkers (carrying their own urban microclimates with them under parasols and thick woollen clothes) viewed the hills. Or they might turn their back to the hills, viewing them instead in a Claude glass – a mirror that framed the view neatly, just as it might appear in their drawing rooms back home. They didn't really want the landscape itself: they wanted a sensation that the landscape might help them to have. They could only stomach an ordered, manicured, controlled pastiche. Many of Romanticism's wolves were neutered.

Denial of the natural world wasn't just a religious meme. John Locke, in his essay *Concerning Human Understanding* (1690), urged his readers to disregard 'all those areas of the natural world which have no connection with man's social and ethical conduct', and in 1775 Samuel Johnson, who should have known better, poured his legendary scorn on those who waxed lyrical about the natural landscape of France. What nonsense, he said: 'A blade of grass is always a blade of grass, whether in one country or another.' No one who has ever *seen* two blades of grass would say anything so stupid.

We can argue about who the first true landscape painter was, and whether things were different outside Europe, but the general position in the visual arts in Europe is clear: from the Neolithic until the European Romantics (and usually thereafter too) nature was, if not actively evil, merely window-dressing for the great human show.

There are some important dissenting voices. They come not from the occasional gentle naturalists but from some of the world's great religions. I include shamanism in that category – not just the shamanism of hunter–gatherers, but the wise women with their cats and their potions, found on the edges of most communities in all ages (and commonly on the top of bonfires and at the bottom of ponds). True hunter–gatherers continued to be a statistically important part of the world's population until fairly recently, and the influence of their nemesis, the nation-state, has been overstated. James C. Scott estimates that until around 1600 CE, much of the world's population had either never seen a routine tax collector or could make themselves fiscally invisible. The hunter–gatherers knew that they were part of the natural world; their whole shape and cast of mind determined by it, dependent on it, subject to it. They knew that each thing was ensouled as they were, that each thing spoke, listened, craved and gave attention and affection, had its own story and was part of a web that joined all to all.

And so did many in the established religions – though, again, sometimes at the edges. Hinduism and Buddhism had always seen as illusory the boundaries between beings, and when Buddhism pushed into Tibet (a classic edge place) it adopted and Buddha-ised the old animism. In the Abrahamic monotheisms nature had a harder time, but it found a home whenever there was serious reflection on the notion that the creation bears the creator's stamp.

Judaism was, and remains, the most suspicious. It feared any confusion between Creator and the created, and always saw the observation of boundaries as central to its mission. The boundaries, it claimed, were established at the Creation. We have seen them already: light / day, land creatures / sea creatures, clean / unclean and so on. It didn't help that – although many of the great festivals of Judaism are arranged around the agricultural year, and at Sukkot Jews are enjoined to live outdoors in transient shelters through which they can see the stars, and so remember their origin as wanderers – Rabbinic Judaism was an essentially urban industry. Part of Israel's big break from its Talmudic past was the birth of a new race of outdoor Jews – Jews who tended orange groves and hiked and fought in the desert. But for many of the world's Jews, old habits die hard. The British Jewish writer Howard Jacobson observes of a Jewish protagonist in one of his novels that 'In the highly improbable event of his being asked to nominate the one most un-Jewish thing he could think of [he] would have been hard pressed to decide between Nature […] and football.'

Nature mysticism in Judaism was left to the kabbalists (as in Islam it was left to the Sufis), and with them took a classically Eastern form: the dissolution of the frontiers of self and ecstatic union with the Other – which included the non-human world. They soaked themselves in Otherness until their own boundaries rotted away.

Christianity was and is rather different. There's a significant split between East and West. The West is, as I have described, inimical to nature – a consequence of an over-emphasis on the transcendence of God and on the afterlife as opposed to this life, and a systematic contempt for matter. This is often expressed in a disdain for sex (a bestial activity) and a failure to remember that St Paul had spoken about the redemption of the whole creation – not just humans.

The Eastern church, despite its ascetic desert power-houses and its regard for monasticism, thinks that matter matters, and has never been embarrassed to accommodate the immanence of God alongside his transcendence. For the Orthodox, God infuses leaves and stoats. The canonical Greek Orthodox morning prayer invokes the Holy Spirit as the 'giver of life, present in all places and filling all things'. 'All things' includes gulls, whales and fungi. 'All places' includes moorlands, rainforests, sub-atomic particles and mitochondria. The druids had their sacred groves, but in Orthodoxy there is no such thing as an un-sacred grove. The Orthodox are unsurprised that Solomon spoke the language of the birds, think that St Francis was more Orthodox than Catholic, and see as their own the Celtic saints who prayed up to their necks in icy water not for penance but to feel one with the water, and who were dried and warmed by tumbling otters when they emerged from the sea. Meditate hard enough on transcendence, says Constantinople, and it becomes immanence too. And vice versa.

And that, more or less, is what I say over the soup.

The Shakespearean broadly agrees with my historical sketch, pointing out that Shakespeare's palpable rapport with uncultivated land was, like everything else about him, precocious. The Professor remarks acidly that if I'd wanted to invite a whole host of straw men to dinner I should have told him so that he could have told the butler to be careful with the candles, and the Physiologist looks uncomfortable, and plays with his bread roll.

"That's quite a lot to say about one bad picture,' continues the Professor. 'But what has it got to do with your project? I thought that you were going to find out what it was like to be a Stone Age hunter–gatherer and a Neolithic settler. Why all this theological mumbo-jumbo?'

It's a fair question. I want to know, I tell them, what sort of creature I am, and so something about what I need in order to thrive. That demands an inquiry into my origins – which also, Professor, happen to be *your* origins. And the sort of creature I am and you are, I've discovered, is a creature who is part of the natural world, not just as a matter of genetic ancestry but as a continuing, defining, everyday fact, however elegant our wine, and however many syllables there are in our words.

The Professor beats his chest and scratches his armpits theatrically.

I appeal to the silent Physiologist, who's pushing a piece of butter round his plate.

'This is no more than Darwin told us, is it? Our cousins are amoeba, which I find flattering and exciting. All this is trite.'

'Well, quite,' breaks in the Professor. 'So why make such a fuss about it? It is indeed trite. We're animals. Our ruling instincts are designed to help us survive and get fat as the hunting and gathering apes that we are. Any of these students' – and he nods at the undergraduates – 'would tell you that. When they go into the City, as, depressingly, most of them will, they'll find themselves in a troop of aggressive male chimps, many of them female, devoted to the establishment of territory, the acquisition of figurative foods and actually and figuratively getting their rocks off. So what's your big news? Why bother to drag yourself and your poor children off into caves?'

'Because I don't trust books,' I reply. 'And anyway you get a wholly different *kind* of knowledge by doing and feeling things.'

The venison arrives, which gives me respite, but no escape.

'You still haven't said why you've gone all supernatural on us,' the Professor pitilessly reminds me once the gravy has done the rounds.

'I haven't said anything at all about the supernatural,' I hear myself saying. 'The experiences *you* choose to call supernatural are completely natural ones. We see them right from the very start of our history as behaviourally modern humans and, if we bother to look, in every moment of our ordinary lives now. If they didn't actually make us human, at least they played an important part in determining the sort of animal we are.'

'Hold on, hold on,' says the Shakespearean, with his mouth full. 'What sort of natural-supernatural stuff are we talking about here?'

I sigh, and start a list: the action of mind at a distance,

including remote sensing and some of what we call clairvoyance; out-of-body experiences and near-death experiences; the presence of mind in non-humans, and the possibility of connection with it; the transgression of what we conventionally understand the rules of time and other dimensions to be, including pre-cognition and the visualisation of usually invisible spatial dimensions; the persistence of personality (whatever that is) after physical death ...

I get no further. The Professor is looking at me, aghast. That merciful waiter is round again, this time to top up the claret.

The Professor's almost lost for words. He fortifies himself with wine, and splutters that he hardly knows where to start. The Shakespearean is leaning back in his chair and looking at me, laconically amused. The Physiologist is dissecting the deer with extraordinary skill and in minute detail.

The Professor takes another swig, wipes his mouth on the napkin and leans forward confidentially.

'I *had* thought,' he says, 'that this was an *Enlightenment* venture. A serious attempt to do something serious. Nothing wrong with fun and games, of course, but *really* ...'

The suave Shakespearean steps in. 'I'd love to hear why you think experiences like that – while no doubt formative for our furry forebears' – he was pleased with that – 'are somehow *real*.'

'Because I've experienced lots of them, that's why,' I say, with as much grace as I can muster. 'And I'm sure you have too.'

'Really?' says the Shakespearean. 'Do tell,' and he beckons over the wine waiter to help me on my way.

And I do. I'm not sure why, except that if someone buys you dinner, you're at their command.

So out it all spills over the rhubarb crumble: spirit foxes; looking down at my own bald head in the hospital; my dead father's dead radio switching on (and Michael Shermer's wedding day radio message from his dead father for good measure); coal tar soap following me round the land; flying in a crow's body over the Derbyshire moorland; a couple of cavemen watching me put on my tie; the grumbling and muttering of a hill, and a wall of mist. And when we head back to the Combination Room, and since all the damage is done, some of the things that are the inheritance of most of us: knowledge that someone's looking at the back of our head; dogs who know that their owners are coming home, even though the owner is hundreds of miles away and has changed their plans at the last minute; *déjà vus*; telephone telepathy; the feeling that the *real* meaning of the world is at hand, but not quite here, and that Plato understood much more than Aristotle; love, intuition, inexplicable feelings of companionability. And surely that will do for the moment, won't it?

'Well,' says the Professor, sitting in an armchair by the fire in the Senior Non-Combination Room later, looking at me through his glass of port, unsure whether to be amused or worried, 'You've had quite a time, haven't you? I hope you don't mind me asking,' – I was fairly sure I would – 'but how can you justify remaining at this university? What you've just described, and what you evidently believe, is completely – *completely* – antithetical to the scientific spirit of this place. I hope to God you don't try to infect any students with these Dark Age fairy tales.'

'Steady on,' intercedes the Shakepearean. 'More things in heaven and earth, and all that. Not to mention free speech.'

'Don't give me that,' thunders the port, which had possessed the Professor's body. 'There are some things that simply *can't* be true. If someone tells me that he can take me to a place where, in base ten, two plus two equals five, do I go with him to see? I do not. I tell him, well … I'm not sure what I tell him.' And he relapses into heavy-breathing silence.

The Shakespearean makes his excuses. 'It's been a very enlightening evening,' he tells me, as he shakes my hand. I can go now too. I bid goodnight to the Physiologist, who has said barely a word all evening, and thank the Professor for his hospitality. 'Welcome,' he grunts, not looking at me, and not getting up. 'I think you know the way out.'

It's raining outside. As I unlock my bike and prepare to cycle off, the Physiologist runs towards me, his gown billowing.

'Have you got a moment?' he pants. 'There's something I want to say.'

'Of course,' I say, and get off my bike. He's embarrassed. Looking at his shoes as he looked as his plate, it floods out.

'It's just this. Something you said rang a bell. About five years ago I was sitting at home when, without any warning, I felt a crushing pain in my chest. I've never had anything like it before or since. It passed off in about five minutes, but then ten minutes later the phone rang. It was my sister, saying that my mother had just collapsed and died in my sister's house. She thought it was a heart attack, and it turned out to be just that. My mother had never had any heart or other problems. She was as fit as a flea. I'd last spoken to her the week before, and she'd been talking excitedly about her plans to go to Paris.'

I don't know what to say, and so say how sorry I am.

'I should have mentioned that back there,' he says. 'I'm sorry I didn't. But you see how it is.'

I say that I don't blame him at all, and ask him if it changed him or his work.

'Not really, to be honest,' he replies. 'You put these things out of your mind, don't you? And it's not as if it changes anything. I work all the time with brains. I see what happens to behaviour when I alter the structure and function of the brain. I know that whatever we mean by mind, it's really the brain.'

This is no time for an argument. We're wet through, and he's told me what he needs to say. I thank him for telling me the story and cycle off home, where X and his son are waiting to see if I've survived my attempt at self-strangulation.

*

I'm tired of evenings like that. I've had so many. Just as the Professor warned, they become assaults by all sides on straw men, and straw men aren't anything like as interesting as real ones. I hate the shrill fundamentalism in me that's elicited by the Professor's shrill fundamentalism: the petulance and counter-petulance, the boring entrenchment and exhausting attrition, the clichés and, most of all, the misrepresentation of the Enlightenment by its defenders.

In this book I've taken a big jump from the Neolithic to the Enlightenment (and often conflated, unhistorically, the Enlightenment, the Renaissance, the Scientific and Industrial Revolutions and the birth and triumph of modern scientism), leapfrogging over state formation; over the extraordinary time around the fifth century BCE, which saw the birth of classical Greek philosophy; over the fecund Judaism of the Second Temple; over Hinduism, Buddhism, Jainism, Confucianism, Zoroastrianism and the great empires of Egypt, Mesopotamia, and China; over Rome; over the advent of Christianity in the first century CE and Islam in the seventh century; over the not-at-all-Dark Ages of Europe; and over the pejoratively named Middle Ages, whose name presumes, revealingly, that *we* are the apex of history.

I've leapfrogged because I'm exploring the notion that we humans are defined by our relationship to the more-than-human-world. Many other relationships changed in the 6,000 or so years between the end of the Neolithic and the sixteenth century, but *that* relationship, although it was refined, qualified, discussed, ignored and abused, didn't change *all* that much. There were two parallel worlds: hunter–gatherers hunted and gathered, seeing themselves as inside nature, talking to it, hearing from it. The rest of the world farmed, seeing itself, to varying degrees, in some ways outside nature, and in any event playing to different rules from those that governed the birds, but subject to the natural world, for better or for worse; needing to bend to it if it couldn't be adequately controlled, needing to propitiate it or, increasingly commonly, its transcendent controllers. There was less conversation with the non-human world: a farmer might bark a command at his dog, or curse his cow, but expected only obedience in return.

Crucially, the world was an organism; personified in some way, either as an entity like Gaia, or a vibrant collection of entities, or a place infused with personhood because it was created by a divine person. It was ensouled. Humans intuited that animals were so-called because at their core was *anima* – a soul. The creation was animate.

And then, in the Scientific Revolution of the sixteenth and seventeenth centuries – a great exorcism began. Souls were driven out of the non-human world, leaving humans (for the time being, and because the church insisted on it) as the only ensouled creatures.

The exorcism began as yet another exercise in line-drawing. Descartes was the man with the pen. He divided reality into two non-communicating realms: the material and the mental. At first this must have seemed innocent: a pedantic piece of philosophical taxonomy. Its outcome was devastating. Mind or soul – call it what you will – was suddenly absent from non-human matter. We can link that absence directly to the ecological outrages of our own age. Killing an ensouled deer or felling an ensouled tree might need some robust moral justification, as all hunter–gatherers know: it is not so obvious that one should agonise over the destruction of a mere machine. For that is what the world and its non-human inhabitants became.

Of course this was not only Descartes's doing. C. S. Lewis, wearing a secular hat, writing generally about the effect of the scientific (and particularly mathematical) reconception of the universe, observes that what 'delivered Nature into our hands' was the use of mathematics to construct hypotheses and the 'controlled observation of phenomena that could be precisely controlled'. This, he says, was to have profound effects on our thoughts and emotions.

> By reducing Nature to her mathematical elements it substituted a mechanical for a genial or animistic conception of the universe. The world was emptied, first of her indwelling spirits, then of her occult sympathies and antipathies, finally of her colours, smells, and tastes. (Kepler at the beginning of his career explained the motion of the planets by their *animae motrices*; before he died, he explained it mechanically).

The immediate result was dualism, not materialism, but dualism was materialism's midwife. Everything other than matter was progressively ignored, and so progressively dismissed as a subject worthy of serious study. The eventual result was the assertion that *nothing other than matter exists at all*. Materialism was never so much a positive doctrine as a failure to pay attention to other categories: it was, and it is, an act of wilful blindness that has hardened into a canonical doctrine that it's dangerous to deny: just ask the poor Physiologist.

The Professor had flashed his Enlightenment Club membership card, and expected masonic reciprocation. When I didn't give it, he felt violated. In his eyes I'd fraudulently procured my way to the dinner table. And he was worried. What was the university coming to if there were people like me within its hallowed walls?

What he meant by the Enlightenment was the eighteenth-century movement that was the culmination of the Scientific Revolution. One of its foremost modern apologists, Steven Pinker, describes it as having four pillars. The central pillar is *reason*. An Enlightenment thinker won't fall back on 'generators of delusion like faith, dogma, revelation, authority, charisma, mysticism, divination, gut feelings, or the hermeneutic parsing of sacred texts'. It doesn't follow from this that humans are believed to be perfectly rational agents.

Then there is *science* and then *humanism*, which provides a secular foundation for morality, using individual human well-being as the touchstone of ethical behaviour. And finally there is intellectual and moral *progress*. There's nothing inevitable about progress, an educated Enlightenment thinker would say, but given the appropriate commitment to reason, science and humanism, it's likely to happen. And, say the apologists, it has. There are glorious jewels glittering in Enlightenment's crown (we're safer, happier, more peaceful, richer, more equal and more democratic, we're told). Well, yes, many things have indeed improved over time, but, as we've noted, correlation is not causation, and I'm suspicious of a historical method that appears to conclude that the Industrial Revolution is an unqualified good and fails to observe that the *imago dei* – the idea that humans are all made in the image of God – is a radically democratic doctrine. The scientific method and, for a while, the more general intellectual culture of the Enlightenment unquestionably produced wonderful benefits. But I don't like being told that it's hypocritical and ungrateful of me to accept dental anaesthesia while at the same time wondering whether humanist ethics scratches every moral itch. It's rather like being told that I can't admire the paintings in the Sistine Chapel unless I swear belief in transubstantiation.

I'm also uneasy about the packaging of these principles. It's too neat to be truly historical. The real sound of the Enlightenment was not the sedate announcement of a programme but the uproar of debate, untrammelled by the presumptions of the previous centuries.

That sound was truly glorious. 'We must have the courage to examine everything, to discuss everything, even to teach anything', wrote the leading Enlightenment thinker Condorcet. 'Have the courage to use your own understanding', urged Kant. 'The maxim of thinking for oneself at all times is enlightenment.' And 'Our age is the age of criticism, to which everything must be subjected.'

I want to live in that age. But it's not the age lauded by Steven Pinker, not the age of the sneering Professor or the quaking Physiologist, worried about his fellowship and his mortgage if he speaks the truth. I don't hear the sound of debate or feel the warmth and thrill of convivial exploration in the modern Enlightenment academies: I hear the sound of catechising, and feel, as I felt over the Professor's dinner, the fingers of the thought police on my collar, and the chill of stifled dissent. 'A major breakthrough of the Scientific Revolution', writes Steven Pinker, 'perhaps its greatest breakthrough – was to refute the intuition that the universe is saturated with purpose.' How can he possibly, using scientific methods, 'refute' that intuition? His isn't a statement of science or reason, but a non-negotiable clause in a religious creed.

In 1981 the biologist Rupert Sheldrake published *A New Science of Life*, which proposed a new mechanism for some commonly observed phenomena, questioning the adequacy of the reigning materialist reductionist paradigm. It produced an extraordinary response. The book was, thundered

Sir John Maddox, the then editor of *Nature*, an 'infuriating tract', which was 'the best candidate for burning there has been for many years'. Maddox, interviewed later about his outburst, was unrepentant. He was 'offended' by Sheldrake's work, and it could be 'condemned in exactly the language that the Pope used to condemn Galileo, and for the same reason. It is heresy.'

My purpose here is not to defend Sheldrake's thesis, but simply to ask: why was Maddox so fearful and upset? I think his explicitly religious language makes it obvious. He saw himself as the custodian of the religious orthodoxy of materialism, and he was worried that Sheldrake was eroding the creed. If the creed were truly the Enlightenment insistence on thoroughgoing scepticism, concerned only with the elucidation of the truth about the natural world, wouldn't he have given Sheldrake a column rather than a roasting? But it's not, and he didn't. For the Enlightenment Taliban, science isn't a method: it's a religion. They're faithful people, and many will cling to their catechism long after it has been shown to be laughable. Conservatism is the easy and lazy option.

Paranoid fundamentalism is the end stage of any movement. When people abandon argument and fall furiously back on shrill assertion, you can be sure that the writing is on the wall. To use Maddox-like language, we're in the End Times of Enlightenment materialism. The old, comforting certainties have been shown to be half-truths. We'll see some examples in a moment.

For people like the Professor this is terrifying. For real scientists it's thrilling.

I'm sorry for many of the biologists I know. When they did their PhDs they were told that, ascending on the ropes fixed in the eighteenth and nineteenth centuries, they would be in the party that triumphantly summited the great mountain of Nothing-Buttery, from which they would have an undistorted view of the entire universe. The natural world, they were solemnly assured, held no mysteries: *everything* could and would be comfortably accommodated in their Enlightenment paradigm of material reductionism.

They were misled. The paradigm is creaking and cracking. Dawkins is an embarrassment. Genetic determinism is dead. We know that genes are in an exciting conversation with the environment, and goodness knows how wide the definition of 'environment' will turn out to be. Lamarck – so long reviled – is back, but he has been renamed 'epigenetics'. Genetics generally doesn't have anything like the explanatory or predictive power it was once thought to have. Genes aren't selfish – or at least not *merely* selfish. In fact, *nothing* is *merely* anything any more.

It's time for biologists to come out: to acknowledge that the power of their old, worn axioms has been overstated, to listen to the creaking of their paradigm and either mend it or get a new one. They are material reductionists from nine to five – for the sake of their salary and their tenure and out of cognitive dissonance. They spend their working days in a virtual reality, built on premises they know to be false. When they get home from the lab, they acknowledge (at least when they read bedtime stories to their children, mourn their dead parents and pat their dogs) that humans and dogs, if not pot plants, are more than machines, that altruism can't be wholly explained by reciprocal altruism or kin selection and that there's more to minds than brain chemistry. They stand wonderingly at sunsets, weep at the St Matthew Passion and think that Wordsworth is wiser than their boss. This queasy oscillation between mutually inconsistent worlds is not a happy way to live. It's time they became whole people, living joined-up lives.

That's the biologists. It is very, very different in mathematics and the physical sciences. There, Enlightenment scepticism is alive and well, and it has produced real scientific progress. It could and should also produce a complete reorientation of our attitude towards all other beings and indeed all other things: precisely the type of moral progress so devoutly sought by the architects of the Enlightenment.

My complaint against the Enlightenment culture I saw over dinner, and that I see in my fearful biologist friends, is that it is not compliant with its own principle of fearless inquiry, and so has become a creed just as tyrannical as the religions against which it railed, and an obstacle to real human and non-human thriving. For you cannot effectively promote the thriving of creatures whose nature you do not understand and whose nature you choose not to understand. Proper scientific scepticism, as seen in the mathematical and physical sciences, suggests that a central core of *all* things is Mind. If that's true, it changes everything. It changes everything, in fact – ontological, ethical and epistemological – to a position close to that of Upper Palaeolithic hunter-gatherers.

To have proper humanist ethics you need to know what a human is. To behave properly towards cows, chickens, mountains and friends, you need to know what they are too.

The Professor believes that all these things are just lumps of matter. They are indeed lumps of matter. Has it been demonstrated that they are *just* lumps of matter? It has not. But the Professor's problem is much greater than that: it is that no one has the first idea what matter is. All we know is how, in some circumstances, it behaves. We have mere metaphors for what it *is*: 'congealed energy', for instance. No remotely foreseeable progress in physics will do anything other than refine the metaphor. We might get more resonantly poetic metaphors, or metaphors that accord slightly less unsatisfactorily with the equations. But metaphors they will remain.

Newton and others saw the natural world as a piece of clockwork. It was implicit in that model that everything in the natural world is in principle explicable and predictable. The splendid successes of the natural sciences and their related technologies hardened that implication into a certainty – into a delusion that the Newtonian model tells the whole story, and accordingly that to apply it will, given enough time, effort and thought, explain and fully describe everything. Science, seen in this way, becomes (at least potential) omniscience.

A bit of realism, humility and Enlightenment rationalism are long overdue. A good place to start the process of getting real, humble and enlightened is the study of consciousness.

The current state of consciousness studies is easily summarised. No one has the faintest idea about the point, nature or location of consciousness. 'Give us time,' plead the biologists. No, sorry. Time's up. You've had 40,000 years or so. Not only have you made no progress, but there is nothing whatever to suggest that, with your dogmatic materialist view of the world, you could, given more time, make any progress, and a great deal to suggest that you couldn't.

Listen: as the Shakespearean said, there are more things in heaven and earth, Professor, than are dreamt of in your Newtonian clockwork philosophy. The clockwork philosophy isn't the full story: it's an approximation, a description of how things, on a big scale, tend to behave. And when you assert that everything can in principle be accommodated within it, you are simply wrong. This is old news. Live with it.

Between 1927 and 1955 (when Einstein died) a debate raged between Einstein and Niels Bohr (the co-author, with Werner Heisenberg, Max Born and others, of the now mainstream Copenhagen account of quantum mechanics). It was about whether *any* theory could, *in principle*, correspond perfectly to the world. Einstein (perhaps surprisingly for the architect of relativity) agreed with the Professor, with all my hopeful biologist friends and with Newton: there could and would eventually emerge a theory that explained and predicted everything. No, said Bohr: uncertainty is a crucial element of the very nature of things; part of the weave of the world. The Heisenberg uncertainty principle means that one cannot speak of the behaviour of particles independently of the process of observation, and the principle of quantum complementarity means that to describe phenomena fully one has to assume behaviour in terms of both particles and waves – yet particle and wave behaviours are mutually exclusive. Heisenberg had written:

> When we speak of the picture of nature in the exact science of our age, we do not mean a picture of nature so much as *a picture of our relationships with nature* [...] Science no longer confronts nature as an objective observer, but sees itself as an actor in this interplay between man and nature. The scientific method of analysing, explaining and classifying has become conscious of its limitations, which arise out of the fact that by its intervention science alters and refashions the object of investigation [...] [M]ethod and object can no longer be separated.

The consciousness of the observer, then, is inextricably entangled with, and affects, whatever is being observed. Heisenberg's assertion that science was 'conscious of its limitations' is, over half a century later, only patchily true.

Einstein and some Princeton colleagues published, in 1935, what they thought was an emphatic demonstration of Bohr's error. If Bohr were correct, they said, the behaviour of two particles that had once interacted with one another would ever thereafter correlate, however far apart they were in space or time. This couldn't be right, since the theory of relativity decreed that nothing could travel faster than the speed of light – and hence two particles could not communicate with one another faster than the speed of light.

It has now been shown empirically, beyond any doubt, that Bohr was right. Once related, always related. No mathematical machete can hack through the quantum entanglement. On the quantum level, space and time seem irrelevant: this is the doctrine of quantum non-locality.

No one has been able to suggest how consciousness could have emerged from unconscious matter – a problem that has led philosophers such as Alfred North Whitehead, Timothy Sprigge, David Griffin, Thomas Nagel, David Chalmers and Galen Strawson to argue for the ancient solution – that matter is not unconscious at all. This makes the poor old biologists come all over funny. But, as I've seen again and again over more congenial High Table dinners than the Professor's, physicists, used to quantum non-locality and entanglement, don't raise an eyebrow.

The consciousness (whatever that is) of matter (whatever that is) seems to be the simplest explanation for the effects anticipated by Heisenberg and Bohr and subsequently demonstrated by audacious experiments in Berkeley, Orsay and Geneva. A human observer's attention involves whatever can be meant by consciousness. It affects the behaviour of something other than the human (if anything is distinct from anything). Isn't it probable that like affects like? That consciousness engages with consciousness? That Mind speaks to Mind?

Non-locality and entanglement are to do with the behaviour of sub-atomic particles, not, immediately, with the relationship between the brain of a dog and the brain of its distant owner, or between me and the eyes of a person staring at the back of my head. But dog brains and human brains and human eyes are made of those particles.

What's more, all of the sub-atomic particles in the universe were once, so the orthodoxy goes, very, very close indeed to one another just 13.8 billion years ago, at the time of the Big Bang. And if that's right, then *every* particle in the universe is entangled with *every other* particle in the universe, mutually affecting behaviour for ever. The oneness spoken of by the mystics may be a fact. If an electron in this evening's lamb chop affects and is affected by an electron in a quasar 15,000 million light years away, it doesn't seem so farcical to think that your dog might be able to pick up your changed intention when you're a hundred miles down the road.

Around the interlocking black-and-white symbols of Yin and Yang is a circle, making them one. In the Sanskrit tradition being (*sat*), consciousness (*chit*) and bliss (*ananda*) are run together as one word: *Satchitananda*. In the Christian tradition the material, sensual Son and the immaterial, omnipresent Spirit are within and relate lovingly to the Father, creator of all, whose character infuses all of the creation.

The Physiologist's insistence, as we stood in the rain, that brain *equalled* Mind, is understandable. For there is obviously *some* relationship between brain states and consciousness. If I get an injection of an anaesthetic drug, or a truck runs over my head, my consciousness will be affected in some way. But to show that *a* is related in some way to *b* falls a long way short (unless you're helped on your way by some illegitimately imported axioms) of showing that *a = b*. There's no shortage of axioms, ripe for import. The Professor has a bookcase full of them in his Tudor rooms.

William James, lecturing at Harvard in 1897, agreed that human consciousness is a function of the brain, but went on to observe that to say that something is a function of the brain is not the same as saying that it is *produced* by the brain. Function can indicate *transmission*: a prism, for instance, refracts light, but does not produce it. The output of a prism looks rather different from the input. The individual characteristics of the prism determine the appearance of the output. Perhaps it's like that with brains and mind. Perhaps your brain gives a particular colour to the part of Mind that beams into it. It doesn't follow that when the truck wheel squashes my brain, so smashing the prism, the individuated part of mind that I called 'me' ceases to exist. It might just relocate.

Brains, then, transmit, mediate and perhaps receive consciousness. They are rather like radios. Damage a radio, and its ability to receive or transmit will be affected. Aldous Huxley spoke of the brain as a 'reducing valve', slowing down the flow of data to levels easily processable by the workaday brain, but which could be slackened off by hallucinogenic drugs – allowing a flood of information at wavelengths that are normally blocked. The number theorist Jason Padgett, widely regarded as a genius who synaesthetically 'sees' fractal patterns in the world, was hopeless at mathematics at high school. He was hit on the head by muggers, and the blows unleashed his clairvoyant relationship with numbers and patterns. It is as if the blows damaged the brain tissue that formed the valve that, in him and us, inhibits the visualisation of fractals. Who's to say how the function of the valve might have changed generally in the population in the last 40,000 years?

We slouch along, usually, in four dimensions: three spatial ones and time. These are the Enlightenment's dimensions, the Professor's dimensions, the biologists' dimensions. But they are not the only ones, as any mathematician will tell you, and they often seem to constrict us. Poets and musicians rail against them; drug users and other pursuers of ecstasy try to transcend them; our recollections of early childhood insist that there are many more than the four. And even as big, sober, grown-ups we often talk as if the four dimensions (and time, in particular) are not our natural habitat. 'Doesn't time fly?' 'It can't *really* be five years since we saw you.' That's not how you talk if you're fully at home in time. 'Do fish complain of the sea for being wet?' asked C. S. Lewis. 'Or if they did, would the fact not strongly suggest that they had not always been, or would not always be, purely aquatic creatures?'

Time makes no sense to us. It is as if we have intuited what the quantum physicists have laboriously demonstrated – which is that time itself is a meaningless category. To begin with, time can't be considered alone, but must be considered as one element in the single medium of space–time. But it goes further than that. We have met the idea of quantum non-locality (in which, across whatever stretches of time and space, previously related entities affect the behaviour of one another instantaneously). If that's right, both space and time shrink into at least irrelevance, if not non-existence. That's not so far from the perspective of Jesus, who, in claiming divinity, confounded the tenses by asserting that 'Before Abraham was, I am.'

There are some curious but common indications that we may sometimes (and perhaps may eventually?) burst out of the straitjacket of our everyday dimensions. Out-of-body experiences, of the sort I had at the hospital, are often accompanied with an apparent multiplication of the dimensions in which we consciously exist. Subjects commonly have 360° vision: 'exactly what one would expect', notes Jeffrey Kripal, 'if a person had suddenly popped into an extra space-time dimension'. Recent work has demonstrated that neuronal networks in the human brain are capable of processing eleven dimensions. We think consciously only of four dimensions. The other four that we have the hardware to inhabit are mathematical abstractions which it would take thick books of equations to explicate and far, far more than a Blake or a Bosch or an El Greco to illustrate. We're wired for *much* more than we usually realise. That's uncharacteristic extravagance on the part of natural selection if we're supposed never to venture out of our usual four dimensions.

I'm sitting in an eating house in Thailand. The sweat's running off my nose and into the bean curd, cockroaches are running over my feet, and the frogs on the riverbank are drowning out Michael Jackson. My notebook's sodden: I might as well have dropped it down the toilet. It's so wet I can't write in it, so I finish off the last of several beers and look at the clock on the wall. It's late at night, but the clock says ten past eight, which is obviously wrong. There's only one other European in the place, a woman who's eating by herself with a copy of *The Snow Leopard* propped up on a soup bowl.

'Excuse me,' I say, 'have you got the right time?'

She looks up from her book and laughs.

'I'm the last person to ask.' She nods at the clock. 'I came in here at ten past eight.'

It is midnight by the time I get back to the dosshouse where I'm staying, because she has told me her story. She's French, and ten years ago she was badly injured when the car she was travelling in was in a head-on collision. She was rushed to hospital, and a haematoma in her brain was drained. She hovered on the edge of life for a while, and had a fairly classic near-death experience, going down a tunnel towards a bright light where her dead relatives were waiting, feeling an overwhelming sense of bliss and peace, and then being tugged reluctantly back to her hospital bed.

Since then, she said, she'd made electronic devices go haywire. Clocks stopped, or sometimes spun wildly. Computers crashed if she went near them. She hadn't yet made a plane plummet from the sky, but she was always worried sick when she flew – or would have been if her near-death experience hadn't banished her fear of death. As we prepared to leave the restaurant, I said I'd pay for her food.

'That's very kind of you,' she said. 'But I bet the card machine won't work if I'm there. I have to carry loads of cash.' She was right: it didn't. She had to stand on the other side of the road before I could pay with the card.

Her hardware, apparently as a result of her accident, seems to be broadcasting *something* at frequencies that are not usually broadcast, to something in clocks, computers and card machines that, because they use the same sort of signals, can pick them up. Consciousness to consciousness? Experiences like that – both the near-death experience and the aftermath – are very common. They are dismissed by the Professor and his ilk who say that they don't happen because they *can't*.

There's plenty of evidence to suggest – as Bohr predicted at the level of quantum phenomena – that Mind acts directly on Mind, producing a peculiar and emphatic kind of knowledge: that the influence of individual minds does not stop at the skull. More systematic investigation is certainly needed. The reason it hasn't happened is, simply, fear that the existing paradigm, on the basis of which all modern scientific careers in the biological and related sciences have been built, might be destroyed or qualified. The studies that have been done in controlled laboratory conditions of phenomena such as telepathy have tended to show small effects – greater than chance, to be sure, but small – in favour of the hypothesis that minds extend beyond heads and operate on matter other than that in the body to which the relevant brain belongs.

Outside the laboratory, though, and in the realm of often well-attested story, the effects are often much more dramatic. There appears to be, as Kripal puts it, a 'privileging of the extreme condition' – by which he means that Mind seems to speak directly to Mind. precisely when other more quotidian types of communication are impossible – typically at the point of death, or in relation to the disclosure (as is common) of future events, or when you're missing your dead father most horribly in a Derbyshire wood. The laboratory experiment necessarily excludes the conditions that are the usual context and justification for direct Mind-to-Mind interaction of the dramatic kind – rather like testing the hypothesis that humans can swim, but only allowing your subjects to try swimming in air. But Mind-to-Mind interaction of a gentler kind is, for most of us, the stuff of everyday experience: knowing what someone is thinking, or is about to say, for instance.

Most of our relationships are based (aren't they?) on things for which there is not and cannot be any demonstrative proof. The Professor's passionate love for his rather imperious wife rests on no empirically or mathematically provable foundation, yet it is far more real and solid than any of the assertions about the nature of society in his published articles, laced with statistics though those assertions are. The substrate of relationship seems to be uncertainty, just as uncertainty (as Bohr anticipated) is one of the load-bearing beams of the universe.

★

I'm on the train, rattling back from London to Oxford, trying and failing to be excited by a novel of unquestionable brilliance, when I see the Professor a few seats away. His headphones are on, his eyes are shut and fluttering in rapture, and his hands are conducting an inaudible orchestra.

I wouldn't dream of entering whatever sacred space he's in, but after a few minutes his hands come down, his eyes open and he sees me, stands up and comes over to sit opposite. We're both uncomfortable about the way we last parted, and relieved to be able to start again.

'So, how are we?' he asks, as if nothing had happened, and we pass a couple of minutes exchanging information that neither of us wants.

'What were you listening to?' I ask.

'Stuff you'd hate,' he spits. He means well, but can't help himself. 'The fruits of high culture. Nothing you can play on an elk-hide drum or a didgeridoo. Bach. All very mathematical and over-educated. Not a whiff of cannabis or semen anywhere.'

And so we're straight back to that dinner, but this time I'm too tired to argue, and I try to turn the conversation away to university politics. Once roused, though, the Professor won't lie down.

'How can you turn your back on the best that humans have achieved?' he wants to know. 'Everything we've understood about ourselves; all the intense intellectual pleasures. Those pleasures, the ones you despise, get me *here*' – he thumps his belly, 'and *here*' – he stabs his finger towards his crotch, 'as well as *here*' – and he slaps his head. '*All* of me. The ego and the id: the left brain and the right.'

And now, suddenly, I start to like the man. Like him very much indeed. But since there are no words to express my complete agreement I nod in a way I hope he'll think is Delphic, and hang on in tense silence until Oxford, looking sagely out of the window.

The Professor's diatribe and my unspoken response together make an unsanitised summary of most of my conversations with my mother in the last few years of her life. She took no prisoners in her war on the wild. Or so it often seemed. Despite her Sicilian roots, and a lot of time spent in the cultural epicentres of the Mediterranean, her skin was polar-white. She feared the sun, I thought, because bright sun meant that she wasn't in her library or a gallery, and that untamed things without end-plates or gilt frames might come to get her. So she taunted me lovingly about my shaggy atavism, and I mocked her, not always gently, about her vertiginously high brow. We drove one another painfully apart, and only when I saw her wrestling with the wolf that was eating her from inside did I understand how much more she knew about wildness than I did – knowledge that she'd got from motherhood, marriage, snotty-nosed schoolkids and the high culture of Europe. She chose to stay inside because she could look after us better that way, and because she understood that if you read Goethe properly you know what it's like to lie spreadeagled on a mountain top, and that to hear Mozart right is to smell a spirit fox and that Andromeda is singing the B Minor Mass. She knew that Apollo and Dionysus are the same god.

I don't recommend her route. It demands a sensibility that makes even the normal sensations of life exquisitely painful. It's far easier, as well as a lot more fun, to run around Derbyshire in a deerskin loincloth. But she showed that cognition, having emerged from the wild, never quite loses touch with it and, if we're careful, and keep cognition on a leash, it can help us to sniff our way back to the place where we were born.

The world is always many steps ahead of us: always baffling: always blinding us with splendour. We need all the help we can get as we limp after it, and that includes the help of partial differential equations, radio telescopes, the Italian Quattrocento, as well as the primary tools – and they must always remain the primary tools – of the mystics and the ecstatics.

If we go into the woods and the rivers and the hills and the seas with *all* this, the wild will feel appreciated. It will know we're trying, and will start to come out and introduce itself. And since you are part of the wild, you should brace yourself for an encounter with *yourself*. It is far more exciting and frightening to be a fully switched-on human than it is to be even the most completely aware badger, otter, fox, deer or swift.

We can have it all! We must have it all! We must be greedy!

Our main way of knowing most of the things that really matter is by the type of knowledge that comes from relationship: of direct encounter unmediated by cognition or language. Once we were adept at that sort of knowledge. We can be again. And we must recruit too, in the effort to know, all the other ways of knowing that we've acquired.

I can't wait to get on with the knowing.

What are we? Dazzling creatures, every electron within us vibrating in unison with and, if we allow it, union with every other electron that the universe contains. The mysterious 'matter' which, for the moment, houses the Mind we call our own, seems to determine some of the ways in which Mind behaves (and may be important in determining the configuration of Mind that you call *you*). But matter seems to be more of a vehicle for Mind – a rather basic trolley – than anything else. Indeed it may be inhibiting its performance, constraining it, clipping its wings.

I look back, embarrassed, to that painfully conflicted, en-mattered barrister, bloodstained on the Scottish hill one afternoon, listening to Schubert that night, then trying to capture human suffering in a syllogism the following week. I wish I could go back and break down the divisions that divided him as Neolithic walls divided the wide wild world. That done, I hope he would have wandered, discovering and living the wild kindness that is everywhere.

★

'Tell me a story,' says Tom.

It is time to tell it yourself.

'*La li-li-li, li-li.*'

EPILOGUE

We're all up in the Derbyshire wood. It has been a high, blue, cold winter day, and now it is a high, black, cold winter night.

We're huddled around a fire. Our shadows are dancing in the treetops. The tick-tocking magpie is on the hawthorn, just above my shoulder, excited to meet the family.

Tom pulls a potato out of the fire and, before eating it, takes it off into the wood for a few minutes. When he comes back, there's a pinch of the potato missing. The others don't notice. He refuses to catch my eye. He can't have missed the moat of coal tar scent that surrounds the fire.

Up by the barn there are two figures I've come to know well. They're standing still, and if I didn't know who they were, I might think they were gateposts. When the wind swoops down from Howden moor and Bleaklow, it ruffles their fur hats. Tom is looking at them too.

I never did find out their real names.

We stamp out the fire and walk up to the house. I turn round and look back down at the wood. The gateposts have moved.

Something or someone is whistling. At first I think it's the wind coming through holes in the drystone wall, but it's not: It's Jonny, our eight-year-old, whistling: *'La li-li-li, li-li.'*

ACKNOWLEDGEMENTS

Massive thanks to the archaeologists and anthropologists who gave so generously of their time, wisdom and coffee, and in particular to: Jan Abbink (Leiden), Justin Barrett (St. Andrews), Vicki Cummings (University of Central Lancashire), Barry Cunliffe (Oxford), Robin Dunbar (Oxford), Avi Faust (Bar Ilan), Israel Finkelstein (Tel Aviv), Clive Gamble (Southampton), Yossi Garfinkel (Hebrew University), the late David Graeber (LSE), Mary MacLeod Rivett (Historic Environment Scotland), Steven Mithen (Reading), Paul Pettitt (Durham), the late Steve Rayner (Oxford), Rick Schulting (Oxford), James C. Scott (Yale), Julian Thomas (Manchester) and Harry Wels (Leiden).

I have the world's finest friends, and they have all helped me to be as human a being as I am, and put up with me when I've fallen short. But I've embroiled some more directly than others in the specific inquiry recorded in this book, and for that I must thank particularly: David Abram, Aharon Barak, Theo Bargiotas, Susan Blackmore, John Butler, Rachel Campbell-Johnston, Stefano Caria, John and Margaret Cooper, James Crowden, Steve Ely, John and Nickie Fletcher, Mariam Motamedi Fraser, Shimon Gibson, Jay Griffiths, David Haskell, Caspar Henderson, Jonathan Herring, Ben Hill, Marie Hauge Jensen, Geoff Johnson, Helen Jukes, Paul Kingsnorth, Marinos Kyriakopolous, Andy Letcher, John Lister-Kaye, Andy McGee, Iain McGilchrist, George Monbiot, Helen Mort, James Mumford, James Orr, Andrew Pinsent, Keith Powell, Jonathan Price, Julian Savulescu, Noam Schimmel, Dietrich Graf von Schweinitz, Stephen Sedley, Karl Segnoe, Martin Shaw, Merlin Sheldrake, Rupert Sheldrake, John Stathatos, Peter Thonemann, Chris Thouless, Colin Tudge, Michael Umney, Emily Watt, Ruth West and Theodore Zeldin.

Manolis Basis, Greece's best bouzouki player, took me inside music as no one else has ever done, and James Bell and all at the Bastard English Session at the Isis Farmhouse Tavern channel long-dead farm labourers every month.

I learned the basics of Mongolian overtone chanting from the wonderful

Jill Purce, and so discovered how literally resonant and fundamentally musical my body is. That experience crucially informed my thinking about the relationship between music and language.

Fran and Kevin Blockley did the best possible job of convincing me that the Neolithic was a Good Thing.

John Lord, the doyen of flint-knappers, kindly and patiently taught me and the children the rudiments of axe- and arrowhead-making, and introduced us to many other prehistoric technologies. He lives in the Stone Ages like no one else I know – not by multiplying anachronisms but by living by the rules of dignity and grave courtesy taught by rock and place.

Saccidananda Ashram in Tamil Nadu narrowed the gap between East and West so that I didn't feel so queasy when I looked into it.

Several monks on Mount Athos, Father Ian Graham of the Greek Orthodox community in Oxford, and my partner in the study of the Talmud, the late Micky Weingarten, taught me that transcendence and immanence are not opponents.

Peter Thonemann read the manuscript in draft and commented with terrifying perspicacity.

Green Templeton College, Oxford, is the greenest and loveliest grove in all academe. The dysfunctional dinner with Professor Black could never have happened there. My friend and colleague Denise Lievesley, former Principal of the college, has done an extraordinary job in making the college a place in which thoughts like the ones in this book could be uttered fearlessly and deconstructed rigorously, and I salute with gratitude and respect the fellows and students.

I've had wonderful help and kindness along the road from many dear friends, and particularly: Elika Barak, Chris and Suz Beckingham, Andrew and Lucy Billen, Magnus Boyd, Rabbi Eli Brackman and Freidy Brackman, Zoe Broughton, Marnie Buchanan, Peter and Laura Carew, Malcolm and Pip Chisholm, Murray Corke, Colette Dewhurst, Issi and Tal Doron, Melina Dritsaki, Tony and Rose Dyer, Kate Foster, Esti Herskowitz, Tony and Sally Hope, Gill and Barry Howard, Mandy Johnson, Pramod Kumar Joshi, Pat Kaufman, Michael and Abigail Lloyd, Nigel McGilchrist, Jolyon and Clare Mitchell, Penelope Morgan, Bewe Munro, Mike Parker, Nigel and Janet Phillips, Costa Pilavachi, Louise Reynolds, Roland Rosner, Kathy Shock, Claire and Mike Smith, Katherine Stathatou, Sarah Thonemann, Caroline Thouless, Hugh Warwick, Jimmy and Melanie Watt, Mark and Sue West, Rob and Alex Yorke, and Joe Zias.

Acknowledgements

I've known for years that I wanted to write a book illustrated by Geoff Taylor. This was my chance! He understood the book completely on his first reading. His wonderful illustrations expound what I want to say far better than I have done.

My peerless agent, Jessica Woollard, believed in this book from the start, and I'm in awe of her energy and dedication. Enormous thanks to my editors at Profile, Helen Conford and Ed Lake, and at Metropolitan, Riva Hocherman, for adopting this strange, turbulent and hysterically ambitious child, and for kind, skilled and disciplined parenting, without which it would have been released into the world far more monstrous than it actually is. Matthew Taylor's superb copy-editing greatly improved the book, and Lottie Fyfe steered it expertly through production.

I have changed some names, locations and times, stitched together incidents from different places and periods in an attempt to make this piecemeal and erratic adventure into a coherent story. For that reason Tom is sometimes made to play parts that were in fact part of my own childhood.

Despite the changes, some will nevertheless be hurt by their representation here, and for that I am truly sorry.

Most of whatever human-ness I have comes from my family, living and dead. My mother and father insisted (and despite being dead still insist) that the business of being a human was the greatest white-knuckle adventure imaginable, and my current teachers on the crash course in humanity are my wife, Mary, and my children, Lizzie, Sally, Tom, Jamie, Rachel and Jonny. I'm amazed that they keep faith with me, and that they usually even keep their temper. It's astonishing, and I'm very, very grateful.

REFERENCES

Epigraph

p. ix   W. B. Yeats, 'Before the World Was Made', in *The Winding Stair and Other Poems* (1933).

Author's Note (pp. 1–11)

p. 1   'for my purposes': I suspect, though, that a lot more happened in Africa (as opposed to Europe) than has generally been recognised by Eurocentric archaeologists.

p. 2   'no clear boundary between the different eras': A word about genes. Mesolithic rather than Upper Palaeolithic haplotypes may or may not predominate in modern English humans. But nobody denies the behavioural continuity between us and the Upper Palaeolithic (or between the Upper Palaeolithic and the Mesolithic), which is what I'm concerned with here. There is, too, well-demonstrated genetic continuity between the Upper Palaeolithic and the Mesolithic in Europe: see Eppie R. Jones, Gloria Gonzalez Fortes, Sarah Connell, Veronika Siska, Anders Eriksson, Rui Martiniano, Russell L. McLaughlin et al., 'Upper Palaeolithic genomes reveal deep roots of modern Eurasians', *Nature Communications* 6(1) (2015): 1–8.

p. 3   'subdued and truncated aurochs': 'Aurochs' looks odd, but it is both the singular and the plural form of the noun. It comes from the Old High German 'ūrohso'. Many thanks to Lottie Fyfe for educating me.

p. 4   'what has happened since the seventeenth century': Some readers may be surprised that there is so little reference to the relevance to the human story and to our present crisis of the issue of brain lateralisation – and particularly to the work of Iain McGilchrist, most notably in *The Master and His Emissary: The Divided Brain and the Making of the Western World* (Yale University

Press, 2009). His thesis is that the two cerebral hemispheres have different functions: each facilitates a different type of attention to the world. The left hemisphere is good at a narrow, focused attention. It is fond of filing, and categories, and is very conservative. It doesn't like its categories to be questioned or confounded. Like the computers with which it is so happy, it has an operating system based on a binary view of reality. It is a nerd. The right hemisphere has a more holistic view of the world; it sees context and relationship, and knows that the truth is often found in paradox. It can live with contradiction. It does not confuse data collection with wisdom. The left hemisphere is supposed to be the brain's administrator: to deal with the quotidian; to keep things tidy so that the whole brain can work optimally. But (goes the thesis) the administrator has progressively seized control: nuance, reflection, wisdom and possibly human identity and the entire planet are casualties of the coup.

Iain is one of my best friends, and his work has had a profound effect on me. I am sure that his thesis is essentially right. It explains far more about the history of ideas and the nature of our current precarious position than any other analysis I know. But I refer little to his thesis here because he and I are trying to do very different things. He searches systematically for a ruling paradigm. I search unsystematically for a few crumbs of comfort and self-knowledge. I defer to Iain for the details of the struggle between the hemispheres, but I have no doubt that in the Neolithic and the Enlightenment the left hemisphere moved rapidly forward in its march towards hegemony.

p. 5   'behaviourally modern humans': I deal at p. 142 with the suggestion that there may be evidence of Neanderthal religious belief/practice.

Upper Palaeolithic: Winter (pp. 15–85)

p. 15   Sarah Moss, *Ghost Wall* (Granta, 2018), p. 31.

p. 15   Igulikik Inuit hunter to Knud Rasmussen, cited by Joan Halifax in *Shaman: The Wounded Healer* (Thames & Hudson, 1982), p. 6.

p. 15   Alwyn Rees and Brinley Rees, *Celtic Heritage: Ancient Tradition in Ireland and Wales* (Thames & Hudson, 1961), p. 16.

p. 25   'some sorts of killing might be wrong': It does not begin to
follow from this that I deny that non-humans can (for instance)
show empathy – let alone have (for instance) a desire to kill. I
deal later with the relevance of non-human consciousness, and
I argue that it is plain that many non-humans have a sense of 'I',
and hence a sense of 'you'. Here, though, I contend only that the
particular way in which the human 'I' emerged and manifested
coloured the way that human love, empathy and so on emerged
and manifested themselves. And that way had very clear ethical
corollaries.

p. 26   'I've seen no sign of them': I'm grateful to Paul Pettitt for
reassuring me that X's presence in Derbyshire, so far from home,
and so early in human history, is just about plausible.

p. 30   'a lot of deeply embedded language': And even supposing that
there is some way of ablating all this, how can I ever write a
book about it? Book-writing depends on all the software that
I'm fantasising about deinstalling. Yet can't we do something
modestly along these lines? If I can't uninstall myself and then
reinstall consciousness, can't I at least get far enough from
myself to be able to ec-statically reappraise myself? And isn't
it at least plausible, if not quite likely, that if I do that, some
new type of consciousness will trickle in, allowing me to guess
at and describe the sensation of a rush of consciousness into a
previously empty vessel?

But no: language will be the main obstacle to the description
of such a sensation. To say that language prevents us from
perceiving anything at all – that it inhibits any meaningful
conversation with the world – is a bit of an embarrassment
for a writer. This is a book about how hopeless, self-defeating
and deadly books are. Don't read it. Do almost anything else. I
have to hope that by writing on and on, language will smash up
language and something else will happen.

p. 31   'Morphine stops you minding': Remember Lawrence of Arabia
in the film, putting out a burning match with his fingers? 'The
trick, William Potter, is not *minding* that it hurts.'

p. 34   'part of the true story': J. David Lewis-Williams and Thomas
A. Dowson, *Images of Power: Understanding Bushman Rock Art*
(Southern Book Publishers, 1989); Jean Clottes and J. David

References

Lewis-Williams, *The Shamans of Prehistory: Trance and Magic in the Painted Caves* (Harry N. Abrams, 1998); David J. Lewis-Williams and David G. Pearce, *San Spirituality: Roots, Expression, and Social Consequences* (AltaMira Press, 2004); David Lewis-Williams, *The Mind in the Cave: Consciousness and the Origins of Art* (Thames & Hudson, 2011); David Lewis-Williams, *Conceiving God: The Cognitive Origin and Evolution of Religion* (Thames & Hudson, 2011); David Lewis-Williams and David Pearce, *Inside the Neolithic Mind: Consciousness, Cosmos and the Realm of the Gods* (Thames & Hudson, 2011). This view has been strongly criticised, and few today would regard it as a complete explanation for prehistoric cave art: see, for instance: Grant S. McCall, 'Add shamans and stir? A critical review of the shamanism model of forager rock art production', *Journal of Anthropological Archaeology* 26(2) (2007): 224–33; and Richard Bradley, *Image and Audience: Rethinking Prehistoric Art* (Oxford University Press, 2009).

p. 35   'Pierced figures in many parts of the world show shamanic business': Mircea Eliade, *Shamanism: Archaic Techniques of Ecstasy* (Princeton University Press, 2004); Joan Halifax, *Shamanic Voices: A Survey of Visionary Narratives* (Plume, 1979); Joan Halifax, *Shaman, the Wounded Healer* (Thames & Hudson, 1982).

p. 35   'The cause of consciousness': For a neurobiological account of the origins of consciousness, see Mark Solms, *The Hidden Spring: A Journey to the Source of Consciousness* (Profile, 2021).

p. 37   'He made his stone tools himself': With John Lord, one of the world's most experienced flint-knappers: https://www.flintknapping.co.uk/

p. 38   'his life would have been forfeit': Many hunter-gatherers have taboos against killing (and certainly eating) carnivores. They are illustrated beautiful by Michelle Paver in her *Chronicles of Ancient Darkness* (Orion Children's, 2008–20), and it may be that the Levitical prohibitions on eating carnivores and scavengers look back to such taboos (as well as denoting Yahweh's disapproval of the shedding of blood: remember that the natural order, as originally conceived – and including humans – was vegetarian).

p. 38   'rabbits in Upper Palaeolithic Derbyshire': There is no evidence of rabbits in Upper Palaeolithic Britain, but rabbit bones

337

carbon-dated to the Roman period have been found at Lynford, Norfolk. They were previously thought to have been introduced by the Normans.

p. 41   'every mouthful was of some ensouled creature': See Joan Halifax, *Shaman, The Wounded Healer*, p. 6.

p. 45   'the moon on her back': See *The Hare Book* (The Hare Preservation Trust, 2015).

p. 57   'a tree root can be like a snake': See Derek Hodgson and Paul Pettitt, 'The origins of iconic depictions: a falsifiable model derived from the visual science of Palaeolithic cave art and world rock art', *Cambridge Archaeological Journal* 28(4) (2018): 591–612.

p. 57   'The nerdish, literal left brain would protest': Iain McGilchrist, *The Master and His Emissary: The Divided Brain and the Making of the Western World*.

p. 59   'They can sometimes even *become* younger': Heather J. Weir, Pallas Yao, Frank K. Huynh, Caroline C. Escoubas, Renata L. Goncalves, Kristopher Burkewitz, Raymond Laboy, Matthew D. Hirschey and William B. Mair, 'Dietary restriction and AMPK increase lifespan via mitochondrial network and peroxisome remodeling', *Cell Metabolism* 26(6) (2017): 884–96; Maria M. Mihaylova, Chia-Wei Cheng, Amanda Q. Cao, Surya Tripathi, Miyeko D. Mana, Khristian E. Bauer-Rowe, Monther Abu-Remaileh et al., 'Fasting activates fatty acid oxidation to enhance intestinal stem cell function during homeostasis and aging', *Cell Stem Cell* 22(5) (2018): 769–78; Andrew W. McCracken, Gracie Adams, Laura Hartshorne, Marc Tatar, and Mirre J. P. Simons, 'The hidden costs of dietary restriction: implications for its evolutionary and mechanistic origins', *Science Advances* 6(8) (2020): 3047.

p. 60   'small bits of bone and mouldering meat': It seems that Tom had intuited the practice, ubiquitous in hunter–gatherers, of giving a portion of his food to the wood before beginning to eat. Traces of the practice persist in some of the mainstream religions too. Note, for instance, the Orthodox Jewish practice of setting aside a portion of bread dough 'for Hashem', the Christian practice of tithing and the food offerings left in Hindu and Buddhist temples.

p. 62   'the process of *anamnesis* – unforgetting': This is, of course,

Plato's idea. It has been brilliantly expounded by Alan Garner in 'Achilles in Altjira', in *The Voice that Thunders* (Harvill Press, 1997).

p. 68   'Aeneas' son, Ascanius, was at his side': Virgil, *Aeneid*, Book II.

p. 68   'Wherever the fire was, there the home would be': For a discussion of the role of the hearth in the ancient world, see Larry Siedentop, *Inventing the Individual: The Origins of Western Liberalism* (Penguin, 2015), pp. 10–13.

p. 71   'Wakeful dreaming is an important spiritual discipline in many religions': Andreas Mavromatis, ed., *Hypnagogia: The Unique State of Consciousness between Wakefulness and Sleep* (Routledge, 1987); Sheelah James, 'Similarities and differences between near death experiences and other forms of religious experience', *Modern Believing* 47(4) (2006): 29–40; Adam J. Powell, 'Mind and spirit: hypnagogia and religious experience', *The Lancet Psychiatry* 5(6) (2018): 473–5.

p. 73   'consciousness is ubiquitous in the natural world': Marian Stamp Dawkins, *Why Animals Matter: Animal Consciousness, Animal Welfare, and Human Well-Being* (Oxford University Press, 2012); Donald R. Griffin, *Animal Minds: Beyond Cognition to Consciousness* (University of Chicago Press, 2013); Carl Safina, *Beyond Words: What Animals Think and Feel* (Macmillan, 2015); Timothy Morton, *Humankind: Solidarity with Non-Human People* (Verso Books, 2017); Pierre Le Neindre, Emilie Bernard, Alain Boissy, Xavier Boivin, Ludovic Calandreau, Nicolas Delon, Bertrand Deputte et al., 'Animal consciousness', *EFSA Supporting Publications* 14(4) (2017): 1196E.

p. 74   'The account of the relationship that makes most sense to me is Iain McGilchrist's': *The Matter with Things* (Penguin RandomHouse, forthcoming).

p. 76   'stories can be told only in the winter or in the dark': Alwyn Rees and Brinley Rees, *Celtic Heritage: Ancient Tradition in Ireland and Wales* (Thames & Hudson, 1961), p. 16.

Upper Palaeolithic: Spring (pp. 86–150)

p. 89   'the great masterpieces of Upper Palaeolithic': Altamira, in Spain, is probably the best example.

p. 91   'wild kindness': Jay Griffiths, *Wild: An Elemental Journey* (Penguin, 2008).

p. 98   'work on the Inuit and the hunter–gatherers of the Pacific Midwest': David Wengrow and David Graeber, 'Farewell to the "childhood of man": ritual, seasonality, and the origins of inequality', *Journal of the Royal Anthropological Institute* 21(3) (2015): 597–619; David Graeber and David Wengrow, 'How to change the course of human history', *Eurozine*. Retrieved from https://www. eurozine. com/change-course-human-history (2018).

p. 99   'A typical modern human has fifth-order intentionality': Taken from Robin Dunbar, *The Human Story* (Faber & Faber, 2011), p. 46.

p. 101   'Because we have separated humanity from nature': William Irwin Thompson, *The Time Falling Bodies Take To Light: Mythology, Sexuality and the Origins of Culture* (Palgrave Macmillan, 1996), p. 102 (emphasis added).

p. 105   'no one who's been to a boarding-school is fit for relationship or public office': For discussion of 'Boarding School Syndrome', see Mark Stibbe, *Home at Last* (Malcolm Down Publishing, 2016), and Joy Schaverien, *Boarding School Syndrome: The Psychological Trauma of the 'Privileged' Child* (Routledge, 2015).

p. 106   'The men of the north of England': I don't know where this is from. I didn't compose it myself, and I can't trace it.

p. 106   'That school did the crucial snatching': This wasn't Shrewsbury School. I went there later, and it was very different.

p. 107   'For me and for them the land is full of agency': Terence McKenna rightly observed: 'Nature loves courage. You make the commitment and nature will respond to that commitment by removing impossible obstacles.'

p. 107   'I finished writing a book about the sea': The book is *A Little Brown Sea* (Fair Acre Press, 2022).

p. 112   'the sun is luring them out': It is only very recently that such ideas have come to seem fanciful. Gilbert White, the seventeenth-century naturalist with legendary powers of observation and inference, thought that swallows, rather than migrating, hibernated in the mud at the bottom of ponds.

p. 115   'the size of the frontal lobes': R. Dunbar, 'Why only humans have language', in R. Botha and C. Knight, eds., *The Prehistory of Language* (Oxford University Press, 2009).

p. 115   'Dunbar's number': R. I. M. Dunbar, 'Mind the gap: or why humans aren't just great apes', in R. I. M. Dunbar, Clive Gamble and J. A. J. Gowlett, *Lucy to Language: The Benchmark Papers* (Oxford University Press, 2014), pp. 3–18.

p. 116   'friends you know and who know you': Robin Dunbar, *Grooming, Gossip, and the Evolution of Language* (Harvard University Press, 1998); Robin Dunbar, 'On the evolutionary function of song and dance', in Nicholas Bannon, ed., *Music, Language, and Human Evolution* (Oxford University Press, 2012), pp. 201–14.

p. 117   'laughter, wordless singing/dancing, language and ritual/religion/story': Dunbar postulates that they arrived in that order.

p. 117   'the endorphins of a laughing policeman': Sandra Manninen, Lauri Tuominen, Robin I. Dunbar, Tomi Karjalainen, Jussi Hirvonen, Eveliina Arponen, Riitta Hari, Iiro P. Jääskeläinen, Mikko Sams and Lauri Nummenmaa, 'Social laughter triggers endogenous opioid release in humans', *Journal of Neuroscience* 37(25) (2017): 6125–31.

p. 118   'modern Kalahari hunter–gatherers': Steven Mithen, *The Singing Neanderthals: The Origins of Music, Language, Mind and Body* (Hachette, 2011), pp. 168–9.

p. 119   'information normally hidden in the subconscious': Michael Winkelman, 'Psychointegrator plants: their roles in human culture', in Michael Winkelman and Walter Andritzky, eds., *Consciousness and Health, Yearbook of Cross-Cultural Medicine and Psychotherapy* (Verlag für Wissenschaft und Bildung, 1996), pp. 9–53; Michael Winkelman, *Shamanism: A Biopsychosocial Paradigm of Consciousness and Healing* (ABC-CLIO, 2010).

p. 119   'our conscious lives are really very boring and trivial': This has made me wonder about how we should regard patients in prolonged disorders of consciousness. We sometimes refer to them nastily as 'vegetables'. But perhaps they're having the time of their lives. Perhaps they are more authentically themselves than they have ever been? Perhaps vegetative life is much more satisfying than ours? I have discussed some of the ethical and legal corollaries of these reflections in various places, for instance: 'It is never lawful or ethical to withdraw life-sustaining treatment from patients with prolonged disorders of consciousness', *Journal of Medical Ethics* 45(4) (2019): 265–70; and 'Deal with the real, not the notional patient, and don't ignore important uncertainties', *Journal of Medical Ethics* 45(12) (2019), 800–801.

p. 123   'Their minds seem to have been rigidly compartmentalised': Steven Mithen, *The Prehistory of the Mind: The Cognitive Origins of Art and Science* (Thames & Hudson, 1999).

p. 123   'Hmmmm': Mithen, *The Singing Neanderthals*, numerous references.

p. 124   'panorama of sounds': Mithen, *The Singing Neanderthals*, p. 245. See too Ian Tattersall, 'The material record and the antiquity of language', *Neuroscience & Biobehavioral Reviews* 81 (2017): 247–54; Dan Dediu and Stephen C. Levinson, 'Neanderthal language revisited: not only us', *Current Opinion in Behavioral Sciences* 21 (2018): 49–55; Lou Albessard-Ball and Antoine Balzeau, 'Of tongues and men: a review of morphological evidence for the evolution of language', *Journal of Language Evolution* 3(1) (2018): 79–89; Mercedes Conde-Valverde et al., 'Neanderthals and *Homo sapiens* had similar auditory and speech capacities', *Nat Ecol Evol* (2021). https://doi.org/10.1038/s41559-021-01391-6

p. 125   'the view of many mainstream archaeologists and anthropologists': Clive Gamble, John Gowlett and Robin Dunbar, *Thinking Big: How the Evolution of Social Life Shaped the Human Mind* (Thames & Hudson, 2014).

p. 125   'Most modern humans have fifth-level intentionality': illustrated above at p. 99.

p. 125   'you need to have sixth-order intentionality': Dunbar, 'Mind the gap: or why humans aren't just great apes'; Peter Kinderman, Robin Dunbar and Richard P. Bentall, 'Theory-of-mind deficits and causal attributions', *British Journal of Psychology* 89(2) (1998): 191–204; James Stiller, and Robin I. M. Dunbar, 'Perspective-taking and memory capacity predict social network size', *Social Networks* 29(1) (2007): 93–104.

p. 125   'more likely to be a Shakespeare if you're female': S. Baron-Cohen, 'Empathizing, systemizing, and the extreme male brain theory of autism', *Progress in Brain Research* 186 (2010): 167–75; M.

References

Adenzato, M. Brambilla, R. Manenti et al., 'Gender differences in cognitive Theory of Mind revealed by transcranial direct current stimulation on medial prefrontal cortex', *Scientific Reports* 7(41219) (2017); Anna Cigarini, Julián Vicens and Josep Perelló, 'Gender-based pairings influence cooperative expectations and behaviours', *Scientific Reports* 10(1) (2020): 1–10.

p. 125   'Spoken language was far better': R. I. M. Dunbar, 'Group size, vocal grooming and the origins of language', *Psychonomic Bulletin and Review* 24 (2017): 209–12.

p. 126   'Dunbar's illustration of sixth-order intentionality': Robin Dunbar, 'The social brain hypothesis and its relevance to social psychology', in Joseph P. Forgas, Martie G. Haselton and William von Hippel, eds., *Evolution and the Social Mind: Evolutionary Psychology and Social Cognition* (Psychology Press, 2007): 21–31, at p. 28.

p. 127   'problems of patients with autism, OCD and ADHD': Autism, OCD and ADHD are examples of conditions in which the ruling medical establishment deems wrong the kind of attention that someone gives to the world. It seems to me that anyone who can attend in what is thought to be an appropriate way to the overwhelming mass of jumbled information with which we are bombarded is far more likely to be suffering from serious mental pathology than people with autism/OCD/ADHD. But that's as may be. In any event, if music helps these 'conditions', it may imply that music is a more satisfactory (and possibly an evolutionarily older) object and/or medium of human attention.

p. 128   'Emotions are primary. Cognition is parasitic on them': For discussion of this issue see Jonathan Haidt, *The Righteous Mind: Why Good People Are Divided by Politics and Religion* (Vintage, 2012), and Joshua D. Greene, *Moral Tribes: Emotion, Reason, and the Gap Between Us and Them* (Penguin, 2013).

p. 128   'the rough-and-ready solutions that experience has taught me will work': Jonathan Haidt, *The Happiness Hypothesis: Finding Modern Truth in Ancient Wisdom* (Basic Books, 2006); and Haidt, *The Righteous Mind.*

p. 128   'music has a developmental, if not evolutionary, priority over language': Mithen, *The Singing Neanderthals*, p. 69. Sherlock Holmes agrees:

'Do you remember what Darwin says about music? He claims that the power of producing and appreciating it existed among the human race long before the power of speech was arrived at. Perhaps that is why we are so subtly influenced by it. There are vague memories in our souls of those misty centuries when the world was in its childhood.'

'That's a rather broad idea,' [Watson] remarked.

'One's ideas must be as broad as Nature if they are to interpret Nature,' he answered.'

Arthur Conan Doyle, *A Study in Scarlet* (Ward Lock, 1887)

Darwin dealt with the issue in *The Descent of Man and Selection in Relation to Sex* (John Murray, 1871), and his thoughts are being increasingly disinterred and rehabilitated: see, for instance, Simon Kirby, 'Darwin's musical protolanguage: an increasingly compelling picture', in Patrick Rebuschat, Martin Rohrmeier, John A. Hawkins and Ian Cross, eds., *Language and Music as Cognitive Systems* (Oxford University Press, 2012), pp. 96–102.

p. 128 'the neural networks for language are built upon or replicate those for music': Mithen, *The Singing Neanderthals*, p. 70.

pp. 128–9 'most fluent, … in Hmmmm and in music': Our faces are very eloquent. We are very unusual, for instance, in having white sclera in our eyes, which make for particularly powerful signalling – even in low light.

p. 129 'we're all naturals': Mithen, *The Singing Neanderthals*, p. 89.

p. 129 '206 kinds of birds they recognise are onomatopoeic': Mithen, *The Singing Neanderthals*, p. 169.

p. 130 'Almost all of them said that it was *mal*': Mithen, *The Singing Neanderthals*, p. 170.

p. 130 'they relate in some mysterious but consistent and fundamental way to qualities in the real world': A full discussion would entail an exploration of Chomsky's 'universal grammar' (see Noam Chomsky, *The Architecture of Language* (Oxford University Press, 2000)). I only note that, although his thesis is hugely controversial, it has fairly recently been boosted by some new findings: see Richard Futrell, Kyle Mahowald and Edward Gibson, 'Large-scale evidence of dependency length minimization in 37 languages', *Proceedings of the National Academy of Sciences* 112(33) (2015): 10336–41.

References

p. 141   'parts of the rock that were later incorporated into the paintings': No doubt bodies were symbolically adorned before any cave was.

p. 141   'Michelle Paver's great saga of the Mesolithic': *Chronicles of Ancient Darkness.* My children's favourite books.

p. 142   'fist-fights to decide who goes to a funeral': For a comprehensive review of non-human primate behaviour towards the dead, see André Gonçalves and Susana Carvalho, 'Death among primates: a critical review of non-human primate interactions towards their dead and dying', *Biological Reviews* 94(4) (2019): 1502–29. See also James R. Anderson, 'Responses to death and dying: primates and other mammals', *Primates* (2020): 1–7. For discussion of what these findings might mean for the interpretation of early human attitudes to death, see Paul Pettitt and James R. Anderson, 'Primate thanatology and hominoid mortuary archaeology', *Primates* (2019): 1–11.

p. 142   'the Neanderthal dead are ... persistently associated with specific locales': Paul Pettitt, 'Landscapes of the dead: the evolution of human mortuary activity from body to place in Pleistocene Europe', in F. Coward, R. Hosfield, M. Pope and F. Wenban-Smith, eds., *Settlement, Society and Cognition in Human Evolution: Landscape in Mind* (Cambridge University Press, 2015), pp. 258–74. For a comprehensive survey of Neanderthal life, death and cognition, see Rebecca Wragg-Sykes, *Kindred: Neanderthal Life, Love, Death and Art* (Bloomsbury, 2020).

p. 143   For a review of some studies of near-death experiences, see Pim van Lommel, 'Near-Death Experiences during cardiac arrest' (2021), https://www.essentiafoundation.org/reading/near-death-experiences-during-cardiac-arrest/. Van Lommel concludes: 'Studies on NDEs seem to suggest that our consciousness does not reside in our brain and is not limited to our brain, because our consciousness has nonlocal properties.'

p. 143   'early hominins tended to believe that minds could survive the death of bodies': The consensus is well summarised by the archaeologist Paul Pettitt: 'Cognitive scientists researching the origins of religion appear to agree that at a relatively early evolutionary stage hominins came to be cognitively predisposed

towards the belief that minds could survive beyond physical
death': Pettitt, 'Landscapes of the dead', p. 262.

p. 143   'For most humans death is not conceived of': Pettitt,
'Landscapes of the dead', p. 262.

p. 144   'to believe in gods, spirits': Pettitt, 'Landscapes of the Dead',
p. 263; and see Paul Bloom, 'Religion is natural', *Developmental
Science* 10(1) (2007): 147–51, at p. 148.

p. 144   '"What are the dead … if they are not symbols?"': Pettitt,
'Landscapes of the dead', p. 258.

p. 144   'a long-dead radio': Michael Shermer, 'Infrequencies', *Scientific
American* 311(4) (2014): 97; discussed in Jeffrey J. Kripal, *The Flip:
Who You Really Are and Why It Matters* (Penguin, 2019), pp. 83–4.

p. 145   'everything humans do is necessarily votive': Alan Garner
wrote: 'Achilles can walk in Altjira [the Australian aboriginal
'Dreaming']. Indeed he must: he has such a lot to remember.
Not least of the memories is that to live as a human being is
in itself a religious act.' 'Achilles in Altjira', in *The Voice that
Thunders*, p. 58. But for him, 'religion' is 'that area of human
concern for, and involvement with, the question of our being
within the cosmos' (p. 55).

Upper Palaeolithic: Summer (pp. 151–74)

p. 151   'Richard Lee calculated': Bruce Chatwin, *The Songlines* (Picador,
1987).

p. 152   'people were stories, and stories grew like fungi out of the soil': I
hope it doesn't need to be said that 'blood and soil' philosophies
are radically misconceived and evil.

p. 155   'As Iain McGilchrist puts it: there *are* no things': McGilchrist, *The
Matter with Things*.

p. 155   'The Bushmen, who walk distances across the Kalahari' and
'Sluggish and sedentary peoples': Chatwin, *The Songlines*.

p. 158   'a thing has to have eyelids to have a soul': Jonathan Balcombe
truly said that if fish had eyelids, rather than relying on water
to bathe their eyes, we wouldn't behave so psychopathically
towards them: Jonathan Balcombe, *What a Fish Knows: The Inner
Lives of Our Underwater Cousins* (Scientific American/Farrar,
Straus and Giroux, 2016).

p. 161   'The Tao that can be spoken': The context is: 'The Tao that can be told is not the eternal Tao. The name that can be named is not the eternal name. The nameless is the beginning of heaven and earth. The named is the mother of ten thousand things. Ever desireless, one can see the mystery. Ever desiring, one can see the manifestations. These two spring from the same source but differ in name; this appears as darkness. Darkness within darkness. The gate to all mystery.' Lao Tzu, *Tao Te Ching*, trans. Jane English and Gia-Fu Fang (Vintage, 1997).

p. 161   'St Paul ... being flung to the ground': Acts 9: 3–9.

p. 162   'It's possible, though very hard, for adults to have true experiential knowledge': Perhaps (it has been suggested to me) the experience of early motherhood is an important exception to this general rule. I wouldn't be surprised.

p. 162   'I listened and studied hard but understood little': Andrew Harvey, *The Direct Path: Creating a Personal Journey to the Divine Using the World's Spiritual Traditions* (Harmony, 2002), Kindle locus 248.

p. 163   'the Stone Age, not the stoned age': I'm sure I'm not the first person to have thought of this weak joke, but I can't find it anywhere else.

p. 163   'echoed in the ontology of some modern indigenous communities': Stephen David Edwards, 'A psychology of indigenous healing in Southern Africa', *Journal of Psychology in Africa* 21(3) (2011): 335–47; Steve Edwards, 'Some Southern African views on interconnectedness with special reference to indigenous knowledge', *Indilinga African Journal of Indigenous Knowledge Systems* 14(2) (2015): 272–83; Jarrad Reddekop, 'Thinking across worlds: indigenous thought, relational ontology, and the politics of nature; or, if only Nietzsche could meet a yachaj' (2014), *Electronic Thesis and Dissertation Repository* 2082: https://ir.lib.uwo.ca/etd/2082. Although there are doubts about its authenticity, the well-known speech of the Duwanish and Suquamish leader Chief Seattle, said to have been delivered in 1854, summarises some strands of indigenous North American attitudes to the natural world: 'Every part of this earth is sacred to my people. Every shining pine needle, every mist in the dark woods, every clearing, and humming insect is holy in the memory and experience of my people. The sap which courses through the trees carries the memory of the red man [...] We are part of the earth and it is part of us [...] The perfumed flowers are our sisters; the deer, the horse, the great eagle, these are our brothers. The rocky crests, the juices of the meadows, the body heat of the pony, and man – all belong to the same family [...] This shining water that flows in the streams is not just water but the blood of our ancestors [...] The water's murmur is the voice of my father's father [...] all things share the same breath – the beast, the tree, the man [...] What is man without the beasts? If all the beasts were gone, men would die from a great loneliness of spirit, for whatever happens to the beast, soon happens to man. All things are connected [...] The earth does not belong to man; man belongs to the earth. This we know. All things are connected like the blood which unites one family. All things are connected. Whatever befalls the earth befalls the sons of the earth. Man did not weave the web of life, he is merely a strand in it. Whatever he does to the web, he does to himself.'

p. 164   '2,070 species in a 90-metre stretch of Devon hedge': https://hedgerowsurvey.ptes.org/biodiversity

p. 164   'the plants have acoustic lives too': Monica Gagliano, Michael Renton, Nili Duvdevani, Matthew Timmins and Stefano Mancuso, 'Acoustic and magnetic communication in plants: is it possible?', *Plant Signaling & Behavior* 7(10) (2012): 1346–8; Monica Gagliano, Michael Renton, Nili Duvdevani, Matthew Timmins and Stefano Mancuso, 'Out of sight but not out of mind: alternative means of communication in plants', *PLoS One* 7(5) (2012); Monica Gagliano, Stefano Mancuso and Daniel Robert, 'Towards understanding plant bioacoustics', *Trends in Plant Science* 17(6) (2012): 323–5; Monica Gagliano, 'Green symphonies: a call for studies on acoustic communication in plants', *Behavioral Ecology* 24(4) (2013): 789–96; Monica Gagliano, 'In a green frame of mind: perspectives on the behavioural ecology and cognitive nature of plants', *AoB Plants* 7 (2015); Monica Gagliano, Mavra Grimonprez, Martial Depczynski and Michael Renton, 'Tuned in: plant roots use sound to locate water', *Oecologia* 184(1) (2017): 151–60.

p. 164   'a green space affects our mood': see Lucy Jones, *Losing Eden:*

References

Why Our Minds Need the Wild (Allen Lane, 2020) and Isabel Hardman, *The Natural Health Service: How Nature Can Mend Your Mind* (Atlantic, 2020).

pp. 166–7   'the dark-skinned humans who crossed from Africa to Europe about 45,000 years ago': Eppie R. Jones, Gloria Gonzalez-Fortes, Sarah Connell, Veronika Siska, Anders Eriksson, Rui Martiniano, Russell L. McLaughlin et al., 'Upper Palaeolithic genomes reveal deep roots of modern Eurasians', *Nature Communications* 6(1) (2015): 1–8; Qiaomei Fu, Cosimo Posth, Mateja Hajdinjak, Martin Petr, Swapan Mallick, Daniel Fernandes, Anja Furtwängler et al., 'The genetic history of ice age Europe', *Nature* 534(7606) (2016): 200–205.

p. 167   'one wouldn't expect them to be bounded by our conventional dimensions of space and time': This issue is discussed in detail below: see pp. 318–24.

p. 167   'a compelling reason to suppose that quantum phenomena apply only at the level of elementary particles': I discuss this objection later, at pp. 320–21.

p. 169   'firing figurines and clay pellets': The pellets and figurines presumably travelled with the wandering communities. They don't, to my knowledge, have a particular association with obviously votive places.

p. 172   'Maps are the worst example of what symbolising can do': Of course hunter–gatherers have mental maps. But hunter–gatherers tend to be insulated from the hubristic danger of maps by the general attitude of deference to the natural world that is one of their defining characteristics. And often the maps themselves (as, famously, is the case with the Australian Songlines) are in a form that is said to emerge directly from the land itself, rather than being a human creation.

Upper Palaeolithic: Autumn (pp. 175–89)

p. 175   '[I]n Sanskrit, which is the great spiritual language of the world': Betty Sue Flowers, ed., *Joseph Campbell and the Power of Myth* (Doubleday and Co., 1988), p. 120.

p. 175   'Achilles can walk in Altjira': Garner, 'Achilles in Altjira', p. 58.

p. 176   'a living Upper Palaeolithic fossil with Turkish blood': Colin

Renfrew, *Archaeology and Language: The Puzzle of Indo-European Origins* (Cambridge University Press, 1990).

p. 177   'Actors are primary': I'm not going to press the point too far, but this fits neatly with the notion that in the Upper Palaeolithic the self – and hence the self's relationship to other entities – blossomed mightily.

p. 177   'There's no hiding shiftily behind a "'The"'': The Basque language always wants to *name*. One has to name in order to praise properly (though the cynic might point to Adam's naming of the animals in Genesis as the quintessential Neolithic act of control: an act antithetical to praise). The shamans say that the natural world is desperate for our praise. Here is Martin Shaw, describing his reconciliation with the wild: 'The first big move was […] reorganising the detritus of my speech to formulate clear and subtle praise for the denizens I beheld in front of me. Not "the Goddess of the River" but "River Goddess". The moment I squeezed "of the" into the mix, thereby hovered an abstraction, and the fox-woman fled the hunter's hut.' Martin Shaw, *Scatterlings: Getting Claimed in the Age of Amnesia* (White Cloud Press, 2016). He'd have been able to make that move faster if he'd spoken Basque to northern Spain rather than Devonian to Dartmoor.

p. 185   'I know that I don't have to go there for forgiveness, or completion, or anything at all': I know that the forgiveness and grace of the land is because I've now walked over enough places to know and become reconciled to, and be taught and fed by, *Place*. You get to the general through the particular. This, I think, was the really great Upper Palaeolithic discovery. It is a paradox, like everything else worth knowing: By wandering you learn to be at home.

Neolithic: Winter (pp. 193–224)

p. 193   'I am not decrying the profession of accountancy': Alan Garner, 'Aback of beyond', in *The Voice that Thunders*, pp. 19–38, at p. 37.

p. 195   'But before all that came the domestication of fire': The case for fire being the most important factor in determining the future of hominins is brilliantly made by James C. Scott in *Against the*

Grain: A Deep History of the Earliest States (Yale University Press, 2017), pp. 37–42.

p. 197 'the universe always gets its own back': Surely this is now obvious. Gaia has had enough of our presumption, and is holding up at least the yellow, if not the red, card. Climate disruption and epidemics that started as zoonoses are just two of many examples of existential threats that are entirely a consequence of a change from the hunter–gatherer orientation to the Neolithic orientation to the world.

pp. 201–2 'We should be slow to link the (slow) Neolithic revolution': I don't deal in this book with the formation of states, the consequences of state formation or the alternatives. That has been done supremely by James C. Scott in *Against the Grain*. I will just note, though, that anarchism is very poorly understood and systematically misrepresented. In both academic and popular literature it's rare to meet an anarchist who's not a straw man. An excellent example is in P. J. O'Rourke's *Holidays in Hell* (Picador, 1989): 'Civilization is an enormous improvement on the lack thereof   [E]very dorm bull session anarchist should spend an hour in Beirut' (p. xvi).

p. 202 'from the stern dialectic of Dunbar's number': For Dunbar's number, see p. 115 above.

p. 206 'the northern Mesopotamian site of Göbekli Tepe': The map may make it look as though Göbekli Tepe is in eastern Anatolia, but for the purposes of this discussion it is most certainly part of Mesopotamia.

p. 208 'took some of the impressive new grains with them as they went home': Steven Mithen, *After the Ice: A Global Human History, 20,000–5000 BC* (Weidenfeld and Nicolson, 2003), p. 67.

p. 211 'the limbic system that governs awareness and general *aliveness*': Scott, *Against the Grain*, pp. 81, 86.

p. 212 'anatomical features of juveniles that wild creatures lose when they reach reproductive age': Adult domestic dogs, for instance, have comparatively shorter muzzles than wolves – muzzles much more like juveniles. And they continue to bark as adults, just as wolf pups do, but as adult wolves do only very rarely. See Temple Grandin and Mark J. Deesing, eds., *Behavioral Genetics and Animal Science* (Academic Press, 2014), pp. 1–40, and, for a discussion of the issue of neoteny in relation to human attractiveness, V. Swami and A. S. Harris, 'Evolutionary perspectives on physical appearance', in Thomas Cash, ed., *Encyclopedia of Body Image and Human Appearance* (Academic Press, 2012), pp. 404–11.

p. 213   'in three weeks a family could harvest the cereals it needed for the year': Jack R. Harlan, 'A wild wheat harvest in Turkey', *Archaeology* 20(3) (1967): 197–201.

p. 214   'the attitude of hunter–gatherers towards agriculturalists': Scott, *Against the Grain*, pp. 7–10.

p. 216   'The story of these two brothers': Genesis 4: 1–21.

p. 218   'the New Jerusalem that rises at the end of time': Revelation 21.

p. 219   'Son of Man has nowhere to lay his head': Matthew 8: 20.

Neolithic: Spring (pp. 225–52)

p. 225   'Death, be not Proud': John Donne, *Holy Sonnets* (1609).

p. 226   'Persistent Complex Bereavement Disorder': *Diagnostic and Statistical Manual of Mental Disorder*, 5th edn (American Psychiatric Association, 2013).

p. 226   'ate her dead husband's ashes with yoghurt': Julia Blackburn, *Time Song: Searching for Doggerland* (Random House, 2019).

p. 228   'newly dead bodies would be left to decompose on top of the more experienced corpses': Julian Thomas, 'Death, identity and the body in Neolithic Britain', *Journal of the Royal Anthropological Institute* 6(40) (2000): 653–668, at p. 659.

p. 228   'the bones would be stroked, kissed and handed to the young children': This reverencing of the bones of the dead is common in many cultures. In Greece to this day, for instance, bones are often disinterred, washed and placed in a family vault.

p. 231   '[T]he notion of a person enclosed within one skin and containing a soul or a mind': Thomas, 'Death, identity, and the body in Neolithic Britain', pp. 657–8.

p. 232   'some form of equivalence between human and cattle remains': Thomas, 'Death, identity, and the body in Neolithic Britain', p. 662.

p. 232   'the number and variety of self–defining, self–locating relationships reduced still further': The dates when these transitions occurred, and the speed with which they occurred, varied greatly according to geography. The Neolithic started and ended earlier in the Near East than in northern Europe, for instance.

p. 232   'there are Neolithic round mounds too': In the north of England, built over linear chambers which contained many bodies: see Thomas, 'Death, identity, and the body in Neolithic Britain', p. 663.

p. 233   'scattered across the chalk hills of southern England': There's an arguable parallel here to the practice (e.g., in Judaea in the Second Temple period) of collecting the bones of an individual in an ossuary rather than mingling them with those of the ancestors – a marker of individuation which (at any rate in Judaea) is generally taken to indicate a belief in the resurrection of the individual.

p. 233   'And if you can point to a place where your ancestors are': For general discussion of this idea in the Near-Eastern context, see Francesca Stavrakopoulou, *Land of Our Fathers: The Roles of Ancestor Veneration in Biblical Land Claims* (T. & T. Clark, 2010).

p. 233   'legitimacy of a claim depends on the legitimacy of the original possession': Absent some fancy modern legal footwork.

p. 236   'Many of the ecological, political and psychological woes of England can be laid at the door of the enclosures': The poet John Clare was driven to the asylum and the grave by the English enclosures – predominantly agricultural reforms – because they banished him from the natural places that were crucial for him.

p. 239   'There was a young man': Langford Reed, *The Complete Limerick Book* (Jarrolds, 1924).

p. 240   'Alan Garner, drawing on this tradition': *Boneland* (Fourth Estate, 2012), p. 47.

p. 242   'the mysterious ladder-like symbols found on cave walls': See pp. 32–4. for discussion of these and other Upper Palaeolithic symbols.

p. 244   'The appearance of rectilinear structures': Richard Bradley, *The Idea of Order: The Circular Archetype in Prehistoric Europe* (Oxford University Press, 2012), pp. 7–11.

p. 245   'if an angular house is to survive anywhere windy, it has to be aligned carefully': Bradley, *The Idea of Order*, p. 29.

p. 246   'Farms and houses, like lines, tunnel vision and constrict the inhabited world': I do not deal here with art other than architecture. Much of the art of the Neolithic and the Bronze Age is, of course, curvilinear. There are exceptions: note, for instance, that in late Neolithic Britain and Ireland, when houses and monuments were round, pottery is often decorated with angular patterns. But the usual rule for the Neolithic is: rectilinear houses, curvilinear art. This dissonance is explored in detail by Richard Bradley in *The Idea of Order*, and it does not confound the thesis I am advancing here.

p. 247 'the ruling metanarrative differed from the one that dictated everyday conduct': Jacques Cauvin, *The Birth of the Gods and the Origins of Agriculture* (Cambridge: Cambridge University Press, 2000). The argument is summarised by Richard Bradley in *The Idea of Order*, pp. 48 and 67. See too on this point, William Irwin Thompson, who in *The Time Falling Bodies Take to Light: Mythology, Sexuality and the Origins of Culture* (Palgrave Macmillan, 1996) contends that the art of the Upper Palaeolithic is characterised by female forms, and that in subsequent eras the rounded female form, so familiar from the Upper Palaeolithic, doesn't cease to dominate. Hence Robert Graves's White Goddess lives on, continuing to rule covertly, in the architecture of temples and other religious buildings and monuments. See Robert Graves, *The White Goddess* (Faber & Faber, 1948).

p. 248 'just one part of a massive complex': For a review of Stonehenge's wider context, see Matt Leivers, 'The Army Basing Programme, Stonehenge and the Emergence of the Sacred Landscape of Wessex', *Internet Archaeology*, 56 (2021).

p. 249 'That's what some of them did': As shown by isotope studies, which can reveal the origin of some artefacts.

p. 249 'Probably they came to lose the fear of death': The fear of death is perhaps *the* great human preoccupation (though I doubt it's a preoccupation just of humans). Unlike many, I don't think that it can be seen as the whole business of religion, or even of those religions which have a clear idea of an afterlife. There are many examples of systems that aim to dissolve the fear without insisting on any metaphysical corollaries. In the ancient world I suspect that the Eleusinian Mysteries is one. Although it was woven round the myth of Demeter and Persephone, I'd be surprised if assent to any propositions in relation to that myth was regarded by anyone as central to the work that the Mysteries did for the initiates.

In our world, psychoanalysis and the burgeoning use of entheogens are obvious examples.

p. 249   'even if unification wasn't the intention, it would inevitably have resulted from such big gatherings': For a comprehensive discussion of the likely purposes of Stonehenge, see Mike Parker Pearson, *Stonehenge: Exploring the Greatest Stone Age Mystery* (Simon and Schuster, 2012).

p. 249   'Other wooden henges were deliberately burned down or dug up and removed': For example, Mount Pleasant in Dorset. For discussion of wood henges as a motif of human transience see: Oliver J. T. Harris and Tim Flohr Sørensen. 'Rethinking emotion and material culture', *Archaeological Dialogues* 17(2) (2010): 145; and Caroline Brazier, 'Walking in sacred space', *Self & Society* 41(4) (2014): 7–14.

p. 249   'Just down the River Avon from Woodhenge there's an avenue': The Woodhenge–Stonehenge avenue may originally have been marked with stones. There is a much more obvious avenue at nearby Avebury, marked with stones on each side, and another at Shap, in Cumbria.

p. 249   'St Paul gave the same answer 1,300 years later': 1 Corinthians 15: 42 (Revised Standard Version): 'What is sown is perishable, what is raised is imperishable.'

p. 250   'they wouldn't be so scared when they had to do it for real': Metaphorical deaths are common in the world's religions. Perhaps the most explicit are in Tibetan Buddhism, where there are systematic attempts to enter the experience of death – and so prepare for it. In Christian baptism the new believer passes through the waters of figurative death, to emerge to the new life beyond.

p. 250   'through a glass, darkly': 1 Corinthians 13: 12 (King James Version): 'For now we see through a glass, darkly; but then face to face: now I know in part; but then shall I know even as also I am known.'

p. 250   'people who appeared not to fear death': Francis Pryor, *Britain*

BC: Life in Britain and Ireland before the Romans (HarperCollins, 2003).

p. 250   'The dark would never win': In many of the old passage tombs from the early Neolithic, at the midwinter solstice when the dark celebrated its greatest triumph over the light, the sun shone on the dead. And in the Stonehenge–Durrington theme park, one of those ephemeral timber circles at Durrington Walls timber circles faced the midwinter light (acknowledging that there is death in the middle of life), while part of the Stonehenge avenue pointed in the direction of the midsummer solstice, the time when the sun is unchallenged (showing that light rules even in the gloomy stone citadel of the dead).

p. 250   'They ruled not just sheep but eternity. It was a big claim': So of course is this claim, but surely it follows from the deliberate juxtaposition of the villages of the living and the dead, and the likelihood of some sort of processional route between them?

p. 252   'safe places for negotiation and fiesta': For discussion see, for example: M. Edmonds, 'Interpreting causewayed enclosures in the past and present', in C. Y. Tilley, ed., *Interpretative Archaeology* (Berg, 1993), pp. 99–142; J. C. Barrett, 'Fields of discourse: reconstituting a social archaeology', in J. Thomas, ed., *Interpretative Archaeology: A Reader* (Leicester University Press, 2000), pp. 23–32; O. J. T. Harris, 'Communities of anxiety: gathering and dwelling at causewayed enclosures in the British Neolithic', in J. Fleisher and N. Norman, eds., *The Archaeology of Anxiety* (Springer, 2016); and Richard Bradley, *The Significance of Monuments: On the Shaping of Human Experience in Neolithic and Bronze Age Europe* (Routledge, 2012).

Neolithic: Summer (pp. 253–76)

p. 253   '[T]he domestication of animals and the breeding of herds': Friedrich Engels, *The Origin of the Family, Private Property and the State* (1884).

p. 254   'with consequent difficulties ... in squeezing out the huge-buttocked calf': L. Fiems, S. Campeneere, W. Caelenbergh and C. Boucqué, 'Relationship between dam and calf characteristics with regard to dystocia in Belgian Blue double-muscled cows', *Animal Science*, 72(2) (2001): 389–94; P. Arthur, 'Double muscling in cows: a review', *Australian Journal of Agricultural Research* 46 (1995): 1493–1515.

p. 254   'causing a lot of maternal pain – and no doubt foetal trauma too – at delivery': Likely to have been achieved by 'flushing' – the practice of giving sheep additional food to promote ovulation.

p. 254   'A modern dairy cow in a conventional unit': Owen Atkinson, *Feeding the Cow*, Webinar Vet, December 2018.

p. 256   'the old Jubilee idea of giving the land and its inhabitants rest and freedom once in a while': See Leviticus 25.

p. 257   'destined for an eternity of torment': An example of this view, from a former (and now disgraced) vicar in a Conservative Evangelical church: 'only today one of our members told me that as a result of being on [a training course] last year it is now impossible for them to walk down the street without being conscious of where most people will spend eternity.' Jonathan Fletcher, *Dear Friends* (Lost Coin Books, 2013), p. 26.

p. 257   'Most of the farmers I know ... are religious, and the ones who insist most loudly that they're not': I have often wondered about this distinction. I presume that to say that one is 'spiritual' is to assert something about your constitution or disposition, and that if you say that you are spiritual but not religious, you are saying that your disposition has not or cannot be constrained by a particular set of theological or metaphysical propositions. The difficulty is that most people who are unequivocally religious but not rankly fundamentalist would say that their spiritual disposition cannot be so constrained either.

p. 258   'Once the direction of prayer was established as vertical': See, for instance, Simon Harrison, 'Cultural efflorescence and political evolution on the Sepik River', *American Ethnologist* 14(3) (1987): 491–507.

p. 259   '150 grass and flower species in all': There may be 150 grass and flower species in a traditional meadow: see http://www.bbc.co.uk/earth/story/20150702-why-meadows-are-worth-saving

p. 260   'Why didn't they opt for freedom and tax-free ease?': I don't deal here with the issue of the earliest states. Some of the issues in state formation are the same as those discussed here, but some are not. The best and most accessible account I know of the birth of the states in the Mesopotamian alluvia is in James C. Scott, *Against the Grain*.

p. 261   'the hunter–gatherers were demographically out-gunned': The post-Neolithic history of the world might coherently be written entirely in terms of the constraints on human relationships imposed by Dunbar's number, but since this book jumps directly from the Neolithic (by the very end of which Dunbar had only just begun to bite) to the Enlightenment, I'm not going to sketch out the outlines of the argument along those lines that I think is possible.

p. 261   'No cattle, no women, no future': Scott, *Against the Grain*, p. 105.

p. 264   'West Country harvesters used to be part-paid in cider': See James Crowden, *Cider: The Forgotten Miracle* (Cyder Press, 1999), p. 15.

p. 264   '"Agricultural lubricant", one Somerset cider maker calls it': Roger Wilkins, of Land's End Farm, Mudgley.

p. 267   'Fifteen billion years ago there was nothing': Chet Raymo, *The Soul of the Night: An Astronomical Pilgrimage* (Cowley Publications, 2005), p. 46.

p. 268   'they'd been told to name all the animals': Genesis 2: 20.

p. 268   'It was a crucial time for Neolithic people': See pp. 248–51 for discussion of the importance of midsummer in the Neolithic.

p. 269   'eaten around 80,000 different animal and plant species': Carolyn Steel, *Sitopia: How Food Can Change the World* (Chatto and Windus, 2020), p. 57.

p. 269   'the staples of three-quarters of the world's population': Steel, *Sitopia*, p. 60.

p. 269   'Just four companies control more than three-quarters': ADM (Archer Daniels Midland), Bunge, Cargill and Dreyfus: see Steel, *Sitopia*, p. 165.

pp. 270–71 'when we settled we became less homicidal': Steven Pinker, *The Better Angels of Our Nature: Why Violence Has Declined* (Viking, 2011), p. 48.

p. 271   'Extrapolating from modern hunter–gatherers to prehistory is dubious': There are, in any event, no living or recent peoples who, in the context I'm discussing here, are usefully analogous to the hunter–gatherers of temperate Europe.

p. 272   'It's the old, old and new, new story': Bruce Chatwin observed in

The Songlines (p. 238) that 'As a general rule of biology, migratory species are less "aggressive" than sedentary ones. There is one obvious reason why this should be so. The migration itself, like the pilgrimage, is the hard journey: a "leveller" on which the "fit" survive and stragglers fall by the wayside. The journey thus pre-empts the need for hierarchies and shows of dominance. The "dictators" of the animal kingdom are those who live in an ambience of plenty. The anarchists, as always, are the gentlemen of the road.'

p. 272   'Africa, Australia and the Arctic indigenous people told Griffiths': All citations here are from Jay Griffiths's talk 'Ferocious Tenderness', at the 'Radical Hope and Cultural Tragedy' conference, 2015: https://www.youtube.com/watch?v=4nzaFmlUDoc. See too Griffiths's *Wild: An Elemental Journey* (Penguin, 2008) and *Kith: The Riddle of the Childscape* (Hamish Hamilton, 2014).

p. 273   'the person who wrote those characters, who created the whole world': The role of writing (and particularly alphabetic – non-pictographic – writing) in the divorce between humans and the non-human world has been extensively and brilliantly explored by David Abram in *The Spell of the Sensuous: Perception and Language in a More-Than-Human World* (Vintage, 1997).

p. 274   'The earliest writing was Sumerian': c.3300 BCE.

p. 274   'ousted by the lines ... of Sumerian cuneiform': By the end of the fourth millennium BCE.

p. 275   'With the alphabet it fell silent': See Abram, *The Spell of the Sensuous*.

Neolithic: Autumn (pp. 277–91)

p. 278   'We shouldn't go to megaliths expecting things to happen': Probably, in the Neolithic, the *building* of megalithic monuments was more important than what happened there.

p. 280   'the desperate focus on the winter solstice': The winter festivals at the big British megalithic monuments seem to have been bigger than those in summer: see e.g. Leivers, referred to in the note to p. 248 above.

p. 280   'a binary world, unimaginable to hunter–gatherers': A satisfactory discussion of binary motifs in ancient and modern thought is well beyond this book. It would have to include, among many other things, an account of structuralism and its critics. I might start with the creation accounts in the book of Genesis (which are perhaps not as binary as they first seem), and then move via Plato, Eastern and Western non-duality, gnosticism, Claude Lévi-Strauss (particularly his book *Le cru et le cuit*, in which he explores the customs of Amazonian tribes) and Jacques Derrida, to computer programming and algorithmic modes of decision-making.

p. 288   'According to *The Lawyer*': From *The Lawyer* (17 January 2020), cited in the Lawtel daily summary.

p. 288   'I try to make them more troubled': In 'Defending the Mysteries', https://vimeo.com/347380878

p. 290   'They didn't give these animals names: they were just "it"': Compare the story of what happened when the Nayaka people of the Nilgiri hills in north-eastern India took up farming. For the first time they started to behave aggressively towards animals, and saw themselves as owners of the newly desouled land: see Danny Naveh and Nurit Bird-David, 'How Persons Become Things: Economic and Epistemological Changes Among Nayaka Hunter-Gatherers', *Journal of the Royal Anthropological Institute*, 20(1) (2014): 74–92.

Enlightenment (pp. 293–329)

p. 293   'When a well-packaged web of lies': Dresden James, widely cited, but origin obscure.

p. 293   'The general materialistic framework of the sciences': Jeffrey J. Kripal, *The Flip*, p. 12.

p. 293   '"There is a mulla-mullung"': Alan Garner, *Strandloper* (Harvill Press, 1996), p. 176.

p. 293   'It's all in Plato, all in Plato': C. S. Lewis, *The Last Battle* (Bodley Head, 1956). More widely available Puffin edition (1964), p. 154.

p. 300   'Aristotle, of course, is an acute and diligent naturalist': See Aristotle, *The History of Animals*, *On the Generation of Animals*, *On the Motion of Animals* and *On the Parts of Animals*.

p. 300   'he identified three types of soul': Aristotle, *On the Soul*.

p. 300   'Lucretius and Pliny the Elder would grace any modern natural history society': See Lucretius' *On the Nature of Things*, and Pliny the Elder's *Natural History*.

p. 300   'Virgil's lush landscape painting in the *Aeneid*': Medieval retellings of the epic of Troy are, like Homer, blind to landscape: see, for instance, the twelfth-century French epic *Roman d'Enéas*. See the discussion of landscape and perspective in Homer, Virgil and medieval literature in Theodore Andersson, *Early Epic Scenery* (Cornell University Press, 1976), and Milton E. Brener, *Vanishing Points: Three Dimensional Perspective in Art and History* (McFarland, 2004).

p. 301   'There was a disdain for untamed nature': Brener, *Vanishing Points*, p. 178.

p. 301   'Natural landscape was dangerous and corrupt': Thomas Burnet, *Sacred Theory of the Earth* (1681), cited in Brener, *Vanishing Points*, p. 179.

p. 301   'like a roaring lion, prowling around, seeking souls to devour': 1 Peter 5: 8.

p. 302   'Or they might turn their back to the hills': For discussion of these practices, see Susan Owens, *Spirit of Place: Artists, Writers and the British Landscape* (Thames & Hudson, 2020).

p. 302   'A blade of grass is always a blade of grass, whether in one country or another': Cited in Brener, *Vanishing Points*, p. 180.

p. 302   'No one who has ever *seen* two blades of grass would say anything so stupid': In relation to England, changing attitudes to landscape have been superbly documented and analysed by Susan Owens in *Spirit of Place*. She notes, for instance, that mountains tended to be seen as terrifying, contends that their sublimity was first recognised by Thomas Gray (1716–1771), of *Elegy in a Country Churchyard* fame, and observes that when doughty Celia Fiennes rode through Derbyshire at the turn of the eighteenth century she appeared to dislike the county mainly because of its relative agricultural uselessness. I differ from Owens on some points: she can see genuine observation of and appreciation of landscape in *Beowulf* (Grendel's lair in particular), and *Gawain and the Green Knight*. I cannot. Indeed it seems to me that both *Beowulf* and *Gawain* are rather good examples of the natural world being seen simply as a backdrop to human drama.

p. 302   'who the first true landscape painter was': Brener contends that it was the German painter Albrecht Altdorfer (*c.*1480–1538).

p. 302   'whether things were different outside Europe': Brener argues convincingly that, although Chinese artists in the medieval period are better and more interested observers of the natural world than European artists, they too had little intrinsic interest in nature, but they did use natural motifs to express truths about themselves: 'The Chinese artists look within themselves and their own spirituality for expression of what they see in nature. The European artists of recent centuries look to the scene itself': Brener, *Vanishing Points*, p. 154.

p. 303   'much of the world's population had either never seen a routine tax collector': Scott, *Against the Grain*, p. 253.

p. 304   'would have been hard pressed to decide between Nature [...] and football': Howard Jacobson, *Coming from Behind* (Vintage, 2003). The protagonist is Sefton Goldberg.

p. 304   'St Paul had spoken about the redemption of the whole creation': Romans 8: 19–22.

p. 310   'big jump from the Neolithic to the Enlightenment': See the discussion in Steven Pinker's *Enlightenment Now: The Case for Reason, Science, Humanism and Progress* (Penguin Random House, 2018), p. 23.

p. 311   'the 6,000 or so years between the end of the Neolithic and the sixteenth century': When the Neolithic ended is, of course, a matter for debate, and depends on where you're talking about.

p. 312   'By reducing Nature': C. S. Lewis, *Poetry and Prose in the Sixteenth Century* (Clarendon Press, 1954), p. 3.

p. 313   'generators of delusion': Pinker, *Enlightenment Now*, p. 8.

p. 313   'many things have indeed improved over time': For examples of the assertions about how things have improved under the benevolent influence of the Enlightenment, see Steven Pinker's books, *The Better Angels of Our Nature: Why Violence Has Declined* and *Enlightenment Now*.

p. 313   'the Industrial Revolution is an unqualified good': In *Enlightenment Now* Pinker writes: 'When the Industrial Revolution released a gusher of useable energy from coal, oil, and falling water, it launched a Great Escape from poverty, disease, hunger, illiteracy, and premature death, first in the West

and increasingly in the rest of the world' (p. 24). No number
of graphs – and indeed no number of metrics of any kind –
are sufficient to establish such a baldly rosy observation. In
considering counter-Enlightenment ideas he deals sharply and
convincingly with the right wing (religious fundamentalism
and nationalism), but less satisfactorily with the left. He says
that 'The left tends to be sympathetic to yet another movement
that subordinates human interests to a transcendent entity, the
ecosystem. The romantic Green movement sees the human
capture of energy not as a way of resisting entropy and
enhancing human flourishing but as a heinous crime against
nature, which will exact a dreadful justice in the form of
resource wars, poisoned air and water, and civilization-ending
climate change' (p. 32). To see human interests as distinct from
those of the non-human world today seems antediluvian and
dangerous. I do not discuss this further in the body of the
text because my concerns are subsumed in the discussion
of the ubiquity of consciousness. But Pinker's advocacy of
a humanistic environmentalism 'more Enlightened than
Romantic, sometimes called ecomodernism or ecopragmatism'
(pp. 121–55) cannot be allowed to pass. He believes that we
can solve the problem of climate change if 'we sustain the
benevolent forces of modernity that have allowed us to solve
problems so far, including societal prosperity, wisely regulated
markets, international governance, and investments in science
and technology' (p. 155). He's proposing that the foxes are put in
charge of the henhouse. That seems unwise. Can the problems
really become the solution? It seems unlikely.

p. 314   'The real sound of the Enlightenment': One of the
Enlightenment's most percipient defenders, Tzvetan Todorov,
observes that it was 'an era of debate rather than consensus'. *In
Defence of the Enlightenment* (London: Atlantic, 2009), p. 9.

p. 314   'We must have the courage to examine everything': Todorov, *In
Defence of the Enlightenment*, p. 44.

p. 314   'The maxim of thinking for oneself': cited in Todorov, *In Defence
of the Enlightenment*, p. 44.

p. 314   'Our age is the age of criticism': cited in Todorov, *In Defence of
the Enlightenment*, p. 35.

p. 314   'A major breakthrough of the Scientific Revolution': Pinker, *Enlightenment Now*, p. 24.

p. 315   'infuriating tract': John Maddox, 'A book for burning?', *Nature* 293(5830) (1981).

p. 315   'For real scientists it's thrilling': A good example of the way that scientific scepticism *should* work is in Michael Shermer's response to the dead radio incident detailed on pp. 144–5. He concluded: 'if we are to take seriously the scientific credo to keep an open mind and remain agnostic when the evidence is indecisive or the riddle unsolved, we should not shut the doors of perception when they may be opened to us to marvel in the mysterious.' Michael Shermer, 'Infrequencies', *Scientific American* 311(4) (2014): 97.

p. 316   'Enlightenment paradigm of material reductionism': A good example of this is in John Maddox's interview about Rupert Sheldrake's book, *A New Science of Life*, discussed above. He said: 'It's unnecessary to introduce magic into the explanation of physical and biological phenomena when in fact there is every likelihood that the continuation of research as it is now practised will indeed fill all the gaps that Sheldrake draws attention to': see https://www.youtube.com/watch?v=QcWOz1xjtsY. Maddox also thought that the Big Bang was 'philosophically unacceptable' and that it would soon be disproved: 'Down with the Big Bang', *Nature* 340(6233) (1898): 425 – another illustration of materialist ideology leading science by the nose.

p. 316   'Lamarck – so long reviled – is back': Lamarck is famous for the promulgation of the notion that evolution proceeded by way of the inheritance of desirable characteristics that are acquired by the parents.

p. 317   'promote the thriving of creatures whose nature you do not understand': Todorov (*In Defence of the Enlightenment*, p. 6) writes about the effect of the Enlightenment on our modern self-definition. If I alter his assertion by the addition of two words (in square brackets here), I think he is right: 'The great upheaval that took place in the three-quarters of a century prior to 1789 is responsible more than anything else for our present-day [lack of] identity.'

p. 317   'We have mere metaphors': Kripal, *The Flip*, p. 55.

References

pp. 318–19   'the now mainstream Copenhagen account of quantum mechanics': There is no universally agreed definition of just what the 'Copenhagen interpretation' of quantum mechanics is. Some would include elements that others would exclude. But there is sufficient agreement about the content of the 'interpretation' to make the expression meaningful, at least for non-specialists. The details of the content do not matter for the purposes of my argument. It is not contentious that core components of the interpretation represented the ruling orthodoxy among quantum physicists in the twentieth century and well into the twenty-first. Most of the central components are still generally agreed, although perhaps there are fewer quantum physicists than there were who would be happy simply to say, without qualification, that they were members of the Copenhagen school.

p. 319   'uncertainty is a crucial element of the very nature of things': Bohr relied on the Heisenberg uncertainty principle (which asserts that the momentum and the position of an electron cannot both be fully known at the same time).

p. 319   'When we speak of the picture of nature': Werner Heisenberg, *The Physicist's Conception of Nature*, trans. Arnold J. Pomerans (Harcourt Brace, 1958), p. 29: original emphasis.

p. 320   'It has now been shown [...] that Bohr was right'. John Clauser et al., University of California, Berkeley; Alain Aspect et al., University of Paris, Orsay; and Nicolas Gisin et al., University of Geneva.

p. 320   'the doctrine of quantum non-locality': For discussion, see Kripal, *The Flip*, pp. 98–103.

p. 320   'a problem that has led philosophers ... to argue for the ancient solution (that matter is not unconscious at all)': Alfred North Whitehead, *Adventures of Ideas* (Macmillan, 1933); Timothy Sprigge, *A Vindication of Absolute Idealism* (Routledge and Kegan Paul, 1983); David Griffin, *Unsnarling the World-Knot: Consciousness, Freedom, and the Mind-Body Problem* (University of Minnesota Press, 1998); Thomas Nagel, 'Panpsychism', in *Mortal Questions* (Cambridge: Cambridge University Press, 1979), pp. 181–95, and *Mind and Cosmos: Why the Materialist Neo-Darwinian Conception of Nature is Almost Certainly False*

(Oxford: Oxford University Press, 2012). The most accessible account of Galen Strawson's view is in 'Realistic materialism: why physicalism entails panpsychism', *Journal of Consciousness Studies* 13(10–11) (2006): 3–31; see too Rupert Sheldrake, 'Is the sun conscious?' *Journal of Consciousness Studies* 28(3–4) (2021): 8–28.

p. 320   'Non-locality and entanglement are to do with the behaviour of sub-atomic particles': We need to be careful about assuming that collections of atomic particles behave in the same way as sub-atomic particles. But hear what the physicist Erich Joos says about this: 'Simply to assume, or rather postulate, that quantum theory is only a theory of micro-objects, whereas in the macroscopic realm per decree (or should I say wishful thinking?) a classical realm has to be valid ... leads to the endlessly discussed paradoxes of quantum theory. These paradoxes only arise because this particular approach is conceptually inconsistent ... In addition, micro- and macro-objects are so strongly dynamically coupled that we do not even know where the boundary between the two supposed realms could possibly be found. For these reasons it seems obvious that there is no boundary.' Erich Joos, 'The Emergence of Classicality from Quantum Theory', in Philip Clayton and Paul Davies, eds., *The Re-Emergence of Emergence: The Emergentist Hypothesis from Science to Religion* (Oxford: Oxford University Press, 2006), pp. 74–5. See too the discussion of this issue by Bernardo Kastrup in *Decoding Jung's Metaphysics: The Archetypal Semantics of an Experiential Universe* (Iff Books, 2021), pp. 46–70.

p. 321   'There's no shortage of axioms, ripe for import': Such as the presumption that we understand what matter is. And that nothing that is not matter exists. And that any phenomenon in the natural world must depend ultimately on matter.

p. 322   'a prism, for instance, refracts light, but does not produce it': See Kripal, *The Flip*, pp. 48–53.

p. 322   'It might just relocate': Perhaps to a more salubrious address? Rumi, adopting Aristotle's view of the hierarchy of souls, observes that when his mineral nature died, he took on a plant soul; when his plant body died, he acquired an animal soul; and when his animal body died, he became a man. Considering his own death, he observes: 'Then what should I fear? I have never become less from dying.' *Masnavi* III: 3901–3906, translated by Ibrahim Gamard, https://www.dar-al-masnavi.org/book3.html

p. 322   'Aldous Huxley spoke of the brain as a "reducing valve"': *The Doors of Perception* (Chatto and Windus, 1954).

p. 322   'the number-theorist Jason Padgett': See Jason Padgett and Maureen Seaberg, *Struck by Genius: How a Brain Injury Made Me a Mathematical Marvel* (Houghton Mifflin Harcourt, 2014).

p. 323   'they had not always been, or would not always be, purely aquatic creatures': Letter to Sheldon Vanauken, 23 December 1950, in Walter Hooper, ed., *The Collected Letters of C. S. Lewis: Narnia, Cambridge and Joy, 1950–1963* (HarperSanFrancisco, 2007).

p. 323   'Before Abraham was, I am': John 8: 58.

p. 323   'exactly what one would expect': Kripal, *The Flip*, pp. 55–6.

pp. 323–4   'neuronal networks in the human brain are capable of processing eleven dimensions': M. W. Reiman et al., 'Cliques of neurons bound into cavities provide a missing link between structure and function', *Frontiers in Computational Neuroscience* (2017).

p. 325   'they don't happen because they *can't*': An excellent example of this attitude is in Gerd Ludemann's discussion of the Ascension. 'As a rule in such a case we did not ask the historical question. In this particular case let me hasten to add that any historical element behind this scene and/or behind Acts 1: 9–11 must be ruled out because there is no such heaven to which Jesus may have been carried.' *The Resurrection: A Historical Inquiry* (Prometheus, 2004), p. 114. James Tabor, commenting on the birth and resurrection of Jesus, holds a similar position: 'Historians are bound by their discipline to work within the parameters of a scientific view of reality. Women do not get pregnant without a male – ever. So Jesus had a human father, whether we can identify him or not. Dead bodies don't rise – not if one is clinically dead – as Jesus surely was after Roman crucifixion and three days in a tomb. So if the tomb was empty the historical conclusion is simple – Jesus' body was moved by someone and likely reburied in another location.' *The Jesus Dynasty* (Harper Element, 2007), pp. 262–3.

p. 325   'the hypothesis that minds extend beyond heads and operate on matter': Etzel Cardeña, 'The experimental evidence for parapsychological phenomena: a review', *American Psychologist* 73.5 (2018): 663; cf. Arthur S. Reber and James E. Alcock, 'Searching for the impossible: parapsychology's elusive quest', *American Psychologist* (2019).

p. 326   'privileging of the extreme condition': Kripal, *The Flip*, pp. 36–7.

p. 326   'uncertainty … is one of the load-bearing beams': See pp. 318–20 above.

SUGGESTED READING

This is not a bibliography. An adequate bibliography would be a list of everything ever written by or about humans. A less adequate bibliography would be a list of everything written about the Upper Palaeolithic, the Neolithic and the Enlightenment. Neither is possible. So this is just what it says – a list of suggestions.

It is quite a long list. If you don't have much time, I'd recommend David Abram, Joseph Campbell, Robin Dunbar, Clive Gamble, Alan Garner, Jay Griffiths, Joan Halifax, Ian Hodder, Timothy Insoll, Paul Kingsnorth, Jeff Kripal, Iain McGilchrist, David Miles, Steven Mithen, Mike Parker Pearson, Paul Pettitt, Steven Pinker, Colin Renfrew, Rick Schulting and Linda Fibiger, James C. Scott and Martin Shaw. No children should be allowed to grow up without reading Michelle Paver.

Abram, David, *The Spell of the Sensuous: Perception and Language in a More-Than Human World* (Vintage, 1997)

Abram, David, *Becoming Animal: An Earthly Cosmology* (Pantheon Books, 2010)

Adams, Cameron, David Luke, Anna Waldstein, David King and Ben Sessa, eds., *Breaking Convention: Essays on Psychedelic Consciousness* (North Atlantic Books, 2014)

Aldhouse-Green, Stephen, and Paul Pettitt, 'Paviland cave: contextualizing the "Red Lady"', *Antiquity* 72 (278) (1998): 756–72

Barham, Larry, Philip Priestley and Adrian Targett, *In Search of Cheddar Man* (Tempus, 1999)

Barrett, Justin L. *Cognitive Science, Religion, and Theology: From Human Minds to Divine Minds* (Templeton Press, 2011)

Bentley Hart, David, *The Experience of God: Being, Consciousness, Bliss* (Yale University Press, 2013)

Blackburn, Julia, *Time Song: Searching for Doggerland* (Random House, 2019)

Blackmore, Susan, *Consciousness: An Introduction* (Routledge, 2013)

Bradley, Richard, *The Past in Prehistoric Societies* (Psychology Press, 2002)

Bradley, Richard, *Image and Audience: Rethinking Prehistoric Art* (Oxford University Press, 2009)

Bradley, Richard, *The Idea of Order: The Circular Archetype in Prehistoric Europe* (Oxford University Press, 2012)

Brener, Milton E., *Vanishing Points: Three Dimensional Perspective in Art and History* (McFarland, 2004)

Broadie, Alexander, *The Scottish Enlightenment* (Birlinn, 2012)

Burns, Jonathan, *The Descent of Madness: Evolutionary Origins of Psychosis and the Social Brain* (Routledge, 2007)

Burroughs, William James, *Climate Change in Prehistory: The End of the Reign of Chaos* (Cambridge University Press, 2005)

Campbell, Joseph, *The Hero with a Thousand Faces* (Pantheon, 1949)

Campbell, Joseph, *The Masks of God*, 4 vols: (1) *Primitive Mythology*; (2) *Oriental Mythology*; (3) *Occidental Mythology*; (4) *Creative Mythology* (Secker & Warburg, 1960)

Campbell, Joseph, *The Way of the Animal Powers*, vol. 1 of *Historical Atlas of World Mythology* (Harper & Row, 1983)

Campbell, Joseph, *The Way of the Seeded Earth*, vol. 2 of *Historical Atlas of World Mythology* (Harper and Row, 1989)

Campbell, Joseph, *The Inner Reaches of Outer Space: Metaphor as Myth and as Religion* (New World Library, 2002)

Campbell, Joseph, and Bill Moyers, *The Power of Myth* (Anchor, 2011)

Cassirer, Ernst, *The Philosophy of the Enlightenment* (Princeton University Press, 1979)

Cauvin, Jacques, *The Birth of the Gods and the Origins of Agriculture* (Cambridge University Press, 2000)

Chatwin, Bruce, *The Songlines* (Picador, 1987)

Chatwin, Bruce, 'The nomadic alternative', in *Anatomy of Restlessness* (Jonathan Cape, 1996), pp. 85–99

Chatwin, Bruce, 'It's a nomad *nomad* world', in *Anatomy of Restlessness* (Jonathan Cape, 1996), pp. 100–106

Clottes, Jean, and David Lewis-Williams, *The Shamans of Prehistory: Trance and Magic in the Painted Caves* (Harry N. Abrams, 1998)

Clutton-Brock, Juliet, *A Natural History of Domesticated Mammals* (Cambridge University Press, 1999)

Clutton-Brock, Juliet, ed., *The Walking Larder: Patterns of Domestication, Pastoralism, and Predation* (Routledge, 2014)

Suggested Reading

Coward, Fiona, Robert Hosfield, Matt Pope and Francis Wenban-Smith, eds., *Settlement, Society and Cognition in Human Evolution* (Cambridge University Press, 2015)

Crockett, Tom, *Stone Age Wisdom: The Healing Principles of Shamanism* (Fair Winds Press, 2003)

Cummings, Vicki, *The Anthropology of Hunter–Gatherers: Key Themes for Archaeologists* (A. & C. Black, 2013)

Cummings, Vicki, *The Neolithic of Britain and Ireland* (Taylor & Francis, 2017)

Cummings, Vicki, and Robert Johnston, eds., *Prehistoric Journeys* (Oxbow Books, 2007)

Cummings, Vicki, Peter Jordan and Marek Zvelebil, eds., *The Oxford Handbook of the Archaeology and Anthropology of Hunter–Gatherers* (Oxford University Press, 2014)

Cunliffe, Barry W., *Europe between the Oceans* (Yale University Press, 2008)

Currie, Gregory, *Arts and Minds* (Oxford University Press, 2004)

Davies, Stephen, *The Artful Species: Aesthetics, Art, and Evolution* (Oxford University Press, 2012)

Dawkins, Marian Stamp, *Why Animals Matter: Animal Consciousness, Animal Welfare, and Human Well-Being* (Oxford University Press, 2012)

Dehaene, Stanislas, *Consciousness and the Brain: Deciphering How the Brain Codes Our Thoughts* (Penguin, 2014)

Dennett, Daniel C., *Consciousness Explained* (Penguin, 1993)

Diamond, Jared M., *Guns, Germs and Steel: A Short History of Everybody for the Last 13,000 Years* (Random House, 1998)

Dossey, Larry, *One Mind* (Hay House, 2013)

Dowd, Marion, and Robert Hensey, *The Archaeology of Darkness* (Oxbow Books, 2016)

Dunbar, Robin, *Grooming, Gossip, and the Evolution of Language* (Harvard University Press, 1998)

Dunbar, Robin, *The Human Story* (Faber & Faber, 2011)

Edmonds, Mark R., and Tim Seaborne, *Prehistory in the Peak* (Tempus, 2001)

Eire, Carlos, *A Very Brief History of Eternity* (Princeton University Press, 2009)

Eisenstein, Charles, *The Ascent of Humanity: Civilization and the Human Sense of Self* (North Atlantic Books, 2013)

Eliade, Mircea, *Shamanism: Archaic Techniques of Ecstasy* (Princeton University Press, 2004)

Engels, Friedrich, *Origin of the Family, Private Property and the State* (1884)

Foer, Jonathan Safran, *Eating Animals* (Penguin, 2010)

Fowler, Chris, *The Archaeology of Personhood: An Anthropological Approach* (Psychology Press, 2004)

Francis, Paul, *The Shamanic Journey* (Paul Francis, 2017)

Gamble, Clive, *The Palaeolithic Societies of Europe* (Cambridge University Press, 1999)

Gamble, Clive, *Origins and Revolutions: Human Identity in Earliest Prehistory* (Cambridge University Press, 2007)

Gamble, Clive, John Gowlett and Robin Dunbar, *Thinking Big: How the Evolution of Social Life Shaped the Human Mind* (Thames & Hudson, 2014)

Garner, Alan, *Strandloper* (Harvill Press, 1996)

Garner, Alan, 'Aback of beyond', in *The Voice that Thunders* (Harvill Press, 1997), pp. 19–38

Garner, Alan, 'Achilles in Altjira', in *The Voice that Thunders* (Harvill Press, 1997), pp. 39– 58

Garner, Alan, *Thursbitch* (Vintage, 2004)

Garner, Alan, *Boneland* (HarperCollins UK, 2012)

Gay, Peter, *The Enlightenment: An Interpretation* (W. W. Norton & Co., 1995)

Gazzaniga, Michael S., *The Consciousness Instinct: Unravelling the Mystery of How the Brain Makes the Mind* (Farrar, Straus and Giroux, 2018)

Goff, Philip, *Consciousness and Fundamental Reality* (Oxford University Press, 2017)

Goldstein, Rebecca, *Betraying Spinoza: The Renegade Jew Who Gave Us Modernity* (New York: Schocken, 2009)

Gosso, Fulvio, and Peter Webster, *The Dream on the Rock: Visions of Prehistory* (SUNY Press, 2013)

Graeber, David, *Bullshit Jobs* (Simon and Schuster, 2018)

Graeber, David, and David Wengrow, 'How to change the course of human history', *Eurozine*. https:/ /www. eurozine. com/change-course-human-history (2018)

Graziano, Michael S. A., *Rethinking Consciousness: A Scientific Theory of Subjective Experience* (W. W. Norton & Co., 2019)

Greene, Joshua D., *Moral Tribes: Emotion, Reason, and the Gap between Us and Them* (Penguin, 2013)

Griffin, Donald R., *Animal Minds: Beyond Cognition to Consciousness* (University of Chicago Press, 2013)

Griffiths, Jay, *Wild: An Elemental Journey* (Penguin, 2008)

Griffiths, Jay, *Kith: The Riddle of the Childscape* (Hamish Hamilton, 2014)

Haidt, Jonathan, *The Happiness Hypothesis: Finding Modern Truth in Ancient Wisdom* (Basic Books, 2006)

Haidt, Jonathan, *The Righteous Mind: Why Good People Are Divided by Politics and Religion* (Vintage, 2012)

Halifax, Joan, *Shamanic Voices: A Survey of Visionary Narratives* (Plume, 1979)

Halifax, Joan, *Shaman, the Wounded Healer* (Thames & Hudson, 1982)

Hamilakis, Yannis, Mark Pluciennik and Sarah Tarlow, eds., *Thinking through the Body: Archaeologies of Corporeality* (Springer Science & Business Media, 2002)

Hampson, Norman, *The Enlightenment* (Penguin, 1990)

Hancock, Graham, *Supernatural: Meetings with the Ancient Teachers of Mankind* (Red Wheel Weiser, 2006)

Hanh, Thich Nhat, John Stanley, David Loy, Mary Evelyn Tucker, John Grim, Wendell Berry, Winona LaDuke et al., *Spiritual Ecology: The Cry of the Earth* (The Golden Sufi Center, 2013)

Hankins, Thomas L., *Science and the Enlightenment* (Cambridge University Press, 1985)

Harner, Michael, *Cave and Cosmos: Shamanic Encounters with Another Reality* (North Atlantic Books, 2013)

Harner, Michael J., Jeffrey Mishlove and Arthur Bloch, *The Way of the Shaman* (New York: HarperSanFrancisco, 1990)

Harvey, Andrew, *The Direct Path: Creating a Personal Journey to the Divine Using the World's Spiritual Traditions* (Harmony, 2002)

Harvey, Graham, and Robert J. Wallis, *Historical Dictionary of Shamanism* (Rowman & Littlefield, 2015)

Herbert, Ruth, *Everyday Music Listening: Absorption, Dissociation and Trancing* (Ashgate, 2013)

Hodder, Ian, *Entangled: An Archaeology of the Relationships between Humans and Things* (John Wiley & Sons, 2012)

Hodder, Ian, ed., *The Meanings of Things: Material Culture and Symbolic Expression* (Routledge, 2013)

Hoffecker, John F., *Modern Humans: Their African Origin and Global Dispersal* (Columbia University Press, 2017)

Hoffman, Donald, *The Case against Reality: Why Evolution Hid the Truth from Our Eyes* (W. W. Norton & Co., 2019)

Huxley, Aldous, *The Doors of Perception* (Chatto and Windus, 1954)

Insoll, Timothy, *Archaeology, Ritual, Religion* (Psychology Press, 2004)

Insoll, Timothy, ed., *The Archaeology of Identities: A Reader* (Routledge, 2007)

Insoll, Timothy, ed., *The Oxford Handbook of the Archaeology of Ritual and Religion* (Oxford University Press, 2011)

Israel, Jonathan Irvine, *Radical Enlightenment: Philosophy and the Making of Modernity, 1650–1750* (Oxford University Press, 2001)

James, William, *The Varieties of Religious Experience: A Study in Human Nature* (Longman, Green & Co, 1902)

Jefferies, Richard, *The Story of My Heart: An Autobiography* (Longman, Green & Co., 1883)

Jones, Andrew, ed., *Prehistoric Europe: Theory and Practice* (John Wiley & Sons, 2008)

Jones, Andrew, *Prehistoric Materialities: Becoming Material in Prehistoric Britain and Ireland* (Oxford University Press, 2012)

Jung, Carl Gustav, *The Earth Has a Soul: C. G. Jung on Nature, Technology and Modern Life* (North Atlantic Books, 2011)

Kalof, Linda, *Looking at Animals in Human History* (Reaktion Books, 2007)

Kastrup, Bernardo, *Decoding Jung's Metaphysics: The Archetypal Semantics of an Experiential Universe* (Iff Books, 2021)

King, Barbara J., *Evolving God: A Provocative View on the Origins of Religion* (University of Chicago Press, 2017)

King, Dave, David Luke, Ben Sessa, Cameron Adams and Aimee Tollan, *Neurotransmissions: Essays on Psychedelics from Breaking Convention* (Strange Attractor Press/MIT Press, 2015)

Kingsnorth, Paul, *Savage Gods* (Little Toller, 2019)

Kingsnorth, Paul, and Dougald Hine, *Uncivilisation: The Dark Mountain Manifesto* (Dark Mountain Project, 2014)

Kripal, Jeffrey J., *The Flip: Who You Really Are and Why It Matters* (Penguin, 2019)

Lanza, Robert, and Bob Berman, *Beyond Biocentrism: Rethinking Time, Space, Consciousness, and the Illusion of Death* (BenBella Books, Inc., 2016)

Lewis, Ioan Myrddin, *Ecstatic Religion: A Study of Shamanism and Spirit Possession* (Psychology Press, 2003)

Lewis-Williams, David, *Conceiving God: The Cognitive Origin and Evolution of Religion* (Thames & Hudson, 2011)

Lewis-Williams, David, *The Mind in the Cave: Consciousness and the Origins of Art* (Thames & Hudson, 2011)

Suggested Reading

Lewis-Williams, David, and Sam Challis, *Deciphering Ancient Minds: The Mystery of San Bushman Rock Art* (Thames & Hudson, 2012)

Lewis-Williams, David, and David G. Pearce, *San Spirituality: Roots, Expression, and Social Consequences* (Rowman Altamira, 2004)

Lewis-Williams, David, and David Pearce, *Inside the Neolithic Mind: Consciousness, Cosmos and the Realm of the Gods* (Thames & Hudson, 2011)

Malafouris, Lambros, *How Things Shape the Mind* (MIT Press, 2013)

Matthiessen, Peter, *The Snow Leopard* (Viking, 1978)

Matthiessen, Peter, *Nine-Headed Dragon River: Zen Journals, 1969–1982* (Shambhala Publications, 1998)

McCarraher, Eugene, *The Enchantments of Mammon: How Capitalism Became the Religion of Modernity* (Belknap, 2019)

McGilchrist, Iain, *The Master and His Emissary: The Divided Brain and the Making of the Western World* (Yale University Press, 2009)

McGilchrist, Iain, *The Matter with Things* (Perspectiva Press, [forthcoming])

McKenna, Terence, *The Archaic Revival* (HarperSanFrancisco, 1991)

McKenna, Terence, *Food of the Gods: The Search for the Original Tree of Knowledge: A Radical History of Plants, Drugs and Human Evolution* (Random House, 1999)

McMahon, Darrin M., *Enemies of the Enlightenment: The French Counter-Enlightenment and the Making of Modernity* (Oxford University Press, 2002)

Miles, David, *The Tale of the Axe: How the Neolithic Revolution Transformed Britain* (Thames & Hudson, 2016)

Mindell, Arnold, *Quantum Mind: The Edge between Physics and Psychology* (Deep Democracy Exchange, 2012)

Mithen, Steven, *The Prehistory of the Mind: The Cognitive Origins of Art and Science* (Thames & Hudson, 1999)

Mithen, Steven, *After the Ice: A Global Human History, 20,000–5000 BC* (Weidenfeld and Nicolson, 2003)

Mithen, Steven, *The Singing Neanderthals: The Origins of Music, Language, Mind and Body* (Hachette, 2011)

Mohen, Jean-Pierre, *Prehistoric Art: The Mythical Birth of Humanity* (Editions Pierre Terrail, 2002)

Monbiot, George, *Feral: Rewilding the Land, the Sea, and Human Life* (University of Chicago Press, 2014)

Morley, Iain, *The Prehistory of Music: Human Evolution, Archaeology, and the Origins of Musicality* (Oxford University Press, 2013)

Morton, Timothy, *Humankind: Solidarity with Non-Human People* (Verso Books, 2017)

Muraresku, Brian C., *The Immortality Key: The Secret History of the Religion with No Name* (St. Martin's Press, 2020)

Neumann, Erich, *The Origins and History of Consciousness* (Routledge, 2015)

Newberg, Andrew, and Eugene G. d'Aquili, *Why God Won't Go Away: Brain Science and the Biology of Belief* (Ballantine Books, 2008)

Outram, Dorinda, *The Enlightenment* (Cambridge University Press, 2019)

Owens, Susan, *Spirit of Place: Artists, Writers and the British Landscape* (Thames & Hudson, 2020)

Pasternak, Charles, ed., *What Makes Us Human?* (Oneworld, 2007)

Paver, Michelle, *Chronicles of Ancient Darkness* (Orion Children's, 2008–21)

Pearson, Mike Parker, *Stonehenge: Exploring the Greatest Stone Age Mystery* (Simon and Schuster, 2012)

Penrose, Roger, Stuart Hameroff and Subhash Kak, eds., *Consciousness and the Universe: Quantum Physics, Evolution, Brain and Mind* (Cosmology Science Publishers, 2011)

Pettitt, Paul, *The Palaeolithic Origins of Human Burial* (Routledge, 2013)

Pettitt, Paul, and Mark White, *The British Palaeolithic: Human Societies at the Edge of the Pleistocene World* (Routledge, 2012)

Pinker, Steven, *The Better Angels of our Nature: Why Violence Has Declined* (Viking, 2011)

Pinker, Steven, *Enlightenment Now: The Case for Reason, Science, Humanism and Progress* (Penguin Random House, 2018)

Plotkin, Bill, *Nature and the Human Soul: Cultivating Wholeness and Community in a Fragmented World* (New World Library, 2010)

Plotkin, Bill, *Wild Mind: A Field Guide to the Human Psyche* (New World Library, 2013)

Price, Neil S., ed., *The Archaeology of Shamanism* (Psychology Press, 2001)

Pryor, Francis, *Farmers in Prehistoric Britain* (Tempus, 1998)

Pryor, Francis, *Britain BC: Life in Britain and Ireland before the Romans* (HarperCollins Publishers, 2003)

Radin, Dean, *Entangled Minds: Extrasensory Experiences in a Quantum Reality* (Simon and Schuster, 2009)

Reill, Peter H., *Vitalizing Nature in the Enlightenment* (University of California Press, 2005)

Suggested Reading

Renfrew, Colin, *Archaeology and Language: The Puzzle of Indo-European Origins* (Cambridge University Press, 1990)

Renfrew, Colin, *The Ancient Mind: Elements of Cognitive Archaeology* (Cambridge University Press, 1994)

Renfrew, Colin, *Prehistory: The Making of the Human Mind* (Modern Library, 2008)

Robb, John, and Oliver J. T. Harris, eds., *The Body in History: Europe from the Palaeolithic to the Future* (Cambridge University Press, 2013)

Roberts, Alice, *Tamed: Ten Species that Changed Our World* (Random House, 2017)

Rosengren, Mats, *Cave Art, Perception and Knowledge* (Springer, 2012)

Rossano, Matt, *Supernatural Selection: How Religion Evolved* (Oxford University Press, 2010)

Russell, Nerissa, *Social Zooarchaeology: Humans and Animals in Prehistory* (Cambridge University Press, 2011)

Safina, Carl, *Beyond Words: What Animals Think and Feel* (Macmillan, 2015)

Schellenberg, Susanna, *The Unity of Perception: Content, Consciousness, Evidence* (Oxford University Press, 2018)

Schulting, Rick J., and Linda Fibiger, eds., *Sticks, Stones, and Broken Bones: Neolithic Violence in a European Perspective* (Oxford University Press, 2012)

Scott, James C., *Seeing Like a State: How Certain Schemes to Improve the Human Condition Have Failed* (Yale University Press, 1998)

Scott, James C., *The Art of Not Being Governed: An Anarchist History of Upland Southeast Asia* (Nus Press, 2010)

Scott, James C., *Against the Grain: A Deep History of the Earliest States* (Yale University Press, 2017)

Sessa, Ben, David Luke, Cameron Adams, Dave King, Aimee Tollan and Nikki Wyrd, *Breaking Convention: Psychedelic Pharmacology for the 21st Century* (Strange Attractor Press, 2017)

Shaw, Martin, *A Branch from the Lightning Tree: Ecstatic Myth and the Grace in Wildness* (White Cloud Press, 2011)

Shaw, Martin, *Scatterlings: Getting Claimed in the Age of Amnesia* (White Cloud Press, 2016)

Shaw, Martin, *Wolf Milk: Chthonic Memory in the Deep Wild* (Cista Mystica, 2019)

Sheldrake, Merlin, *Entangled Life* (Bodley Head, 2020)

Sheldrake, Rupert, *A New Science of Life* (Icon Books, 2005)

Sheldrake, Rupert, *Dogs That Know When Their Owners Are Coming Home, and Other Unexplained Powers of Animals* (Broadway Books, 2011)

Sheldrake, Rupert, *The Presence of the Past: Morphic Resonance and the Habits of Nature* (Icon Books, 2011)

Sheldrake, Rupert, *The Sense of Being Stared at, and Other Aspects of the Extended Mind* (Random House, 2013)

Siedentop, Larry, *Inventing the Individual: The Origins of Western Liberalism* (Penguin Random House, 2015)

Siegel, Daniel J., *Mind: A Journey to the Heart of Being Human*, Norton Series on Interpersonal Neurobiology (W. W. Norton & Co., 2016)

Solms, Mark, *The Hidden Spring: A Journey to the Source of Consciousness* (Profile, 2021)

Stavrakopoulou, Francesca, *Land of Our Fathers: The Roles of Ancestor Veneration in Biblical Land Claims* (T. & T. Clark, 2010)

Steel, Carolyn, *Sitopia: How Food Can Save The World* (Chatto and Windus, 2020)

Talbot, Michael, *The Holographic Universe* (HarperPerennial, 1992)

Tattersall, Ian, *Becoming Human: Evolution and Human Uniqueness* (Houghton Mifflin Harcourt, 1999)

Tattersall, Ian, *The Monkey in the Mirror: Essays on the Science of What Makes Us Human* (Houghton Mifflin Harcourt, 2016)

Thomas, Julian, 'Death, identity and the body in Neolithic Britain', *Journal of the Royal Anthropological Institute* 6(4) (2000): 653–68

Thompson, William Irwin, *The Time Falling Bodies Take To Light: Mythology, Sexuality and the Origins of Culture* (Palgrave Macmillan, 1996)

Todorov, Tzvetan, *In Defence of the Enlightenment* (Atlantic Books, 2009)

Tudge, Colin, *Neanderthals, Bandits and Farmers: How Agriculture Really Began* (Yale University Press, 1999)

Turner, Mark, *The Origin of Ideas: Blending, Creativity, and the Human Spark* (Oxford University Press, 2014)

Vernon, Mark, *A Secret History of Christianity: Jesus, the Last Inkling, and the Evolution of Consciousness* (John Hunt, 2019)

Wallis, Robert J., *Shamans/Neo-Shamans: Ecstasy, Alternative Archaeologies, and Contemporary Pagans* (Psychology Press, 2003)

Wengrow, David, and David Graeber, 'Farewell to the "childhood of man": ritual, seasonality, and the origins of inequality', *Journal of the Royal Anthropological Institute* 21(3) (2015): 597–619

Suggested Reading

Whittle, Alisdair, *The Archaeology of People: Dimensions of Neolithic Life* (Routledge, 2003)

Wittmann, Marc, *Altered States of Consciousness: Experiences out of Time and Self* (MIT Press, 2018)

Wragg-Sykes, Rebecca, *Kindred: Neanderthal Life, Love, Death and Art* (Bloomsbury, 2020)

Wyrd, Nikki, David Luke, Aimee Tollan, Cameron Adams and David King, *Psychedelicacies: More Food for Thought from Breaking Convention* (Strange Attractor/MIT Press, 2019)

Zaidel, Dahlia W., *Neuropsychology of Art: Neurological, Cognitive, and Evolutionary Perspectives* (Psychology Press, 2015)

ABOUT THE AUTHOR

CHARLES FOSTER is the author of *Being a Beast*, which won the 2016 Ig Nobel Award for biology and was a finalist for the Baillie Gifford Prize. He teaches medical law and ethics at the University of Oxford, and his writing has been published in *National Geographic*, the *Guardian*, *Nautilus*, *Slate*, the *Journal of Medical Ethics*, and many other venues. He lives in Oxford, England.